Mathematics:
The Quest for
Truth and Beauty

Problem Solving in Mathematics and Beyond

Print ISSN: 2591-7234
Online ISSN: 2591-7242

Series Editor: Dr. Alfred S. Posamentier
Distinguished Lecturer
New York City College of Technology - City University of New York

There are countless applications that would be considered problem solving in mathematics and beyond. One could even argue that most of mathematics in one way or another involves solving problems. However, this series is intended to be of interest to the general audience with the sole purpose of demonstrating the power and beauty of mathematics through clever problem-solving experiences.

Each of the books will be aimed at the general audience, which implies that the writing level will be such that it will not be engulfed in technical language — rather the language will be simple everyday language so that the focus can remain on the content and not be distracted by unnecessarily sophiscated language. Again, the primary purpose of this series is to approach the topic of mathematics problem-solving in a most appealing and attractive way in order to win more of the general public to appreciate this most important subject rather than to fear it. At the same time we expect that professionals in the scientific community will also find these books attractive, as they will provide many entertaining surprises for the unsuspecting reader.

Published

For the complete list of volumes in this series, please visit www.worldscientific.com/series/psmb

Problem Solving in
Mathematics and Beyond Volume **38**

Mathematics:
The Quest for
Truth and Beauty

James D Stein
California State University, Long Beach, USA

World Scientific

NEW JERSEY · LONDON · SINGAPORE · BEIJING · SHANGHAI · TAIPEI · CHENNAI

Published by

World Scientific Publishing Co. Pte. Ltd.

5 Toh Tuck Link, Singapore 596224

USA office: 27 Warren Street, Suite 401-402, Hackensack, NJ 07601

UK office: 57 Shelton Street, Covent Garden, London WC2H 9HE

Library of Congress Control Number: 2024056953

British Library Cataloguing-in-Publication Data
A catalogue record for this book is available from the British Library.

Problem Solving in Mathematics and Beyond — Vol. 38
MATHEMATICS: THE QUEST FOR TRUTH AND BEAUTY

ISBN 978-981-98-0630-0 (hardcover)
ISBN 978-981-98-0750-5 (paperback)
ISBN 978-981-98-0631-7 (ebook for institutions)
ISBN 978-981-98-0632-4 (ebook for individuals)

For any available supplementary material, please visit
https://www.worldscientific.com/worldscibooks/10.1142/14131#t=suppl

Desk Editors: Kannan Krishnan/Gabriel Rawlinson

Typeset by Stallion Press
Email: enquiries@stallionpress.com

About the Author

James D. Stein received his B.A. in mathematics (1962) from Yale University and an M.A. and Ph.D. from the University of California, Berkeley (1967). His teaching career included 8 years at the University of California, Los Angeles (UCLA), 35 years at California State University, Long Beach (retired in 2013), and 10 years at El Camino Community College in Torrance, California, all in the U.S.

Over his career, he has authored approximately 40 research papers, primarily in pure mathematics, with a few in physics and math education. He has attended and presented at numerous mathematical conferences and has given talks on his research in both the U.S. and Europe, mainly in the 20th century.

Prof. Stein has also written 15 books, most of which are trade books about math and science (published by HarperCollins, Basic Books, McGraw Hill, Wiley, and Princeton University Press). Others include a book of calculus problems (co-authored, National Science Foundation) and one on strategic management (co-authored, Wiley). Two of his books were *Scientific American* Book Club selections, and two have previously been published by World Scientific Press (*The Fate of Schrodinger's Cat* and *Seduced by Mathematics*). His most recent book is *The Milestones of Science*, published in 2023 by Prometheus Books.

As a content expert, he has served as a member of the California Statewide K-12 Textbook Adoption Committee (2000 and 2013) and as a consultant for the Texas Essential Knowledge and Skills program (mid-1990s).

In addition to his research and professional contributions, Prof. Stein has featured in around 40 podcasts (mostly as an interviewer, but several on his own books) for the New Books Network on math and science, including some with high-profile authors. He has given around 200 radio interviews regarding his books and many in-person lectures to book clubs and general gatherings. While his social media presence is limited, several of his recent books have dedicated websites, constructed with the publisher's help.

Contents

Introduction

The dictionary defines a "quest" as a long and arduous search for something. I think most of us, at some point or another, have quibbled over a dictionary definition, and I certainly feel that this definition does the word "quest" a serious injustice. In order to qualify as a quest, IMHO, the "something" has to be a goal with some epic or noble quality about it. The long search by Indiana Jones and his father for the Lost Ark of the Covenant certainly qualifies; trying to find the best place for pizza in your neighborhood doesn't.

Additionally, there has to be passion – or obsession – on the part of those partaking in the quest. Quests have been a part of human history – and drama – for millennia. Some end successfully, as when Jason and the Argonauts successfully retrieved the Golden Fleece, which restored Jason to his rightful place on the throne of Thessaly. Some are unsuccessful and are doomed from the start, such as the quest of Ponce de León to locate the Fountain of Youth. Although maybe this quest wasn't completely unsuccessful, as de León did discover Florida, and many of America's aging population from wintry climates relocate there because, although there is no Fountain of Youth, the weather is at least conducive to a less rapid acceleration of the aging process – or so it is felt.

But quests are also a part of mathematics and science. The scientific quests are generally more well known – the quest to discover the causes and cures for diseases, the nature of evolution, and the structure of the Universe. Many of these quests are still ongoing; it may actually be impossible to come up with final answers. But humanity has benefited greatly from these quests and should continue to do so.

This book, however, focuses on mathematical quests, which have both advantages and disadvantages when compared with other quests. On the positive side of the ledger, they are usually considerably less expensive, less risky (although not always so), and can be pursued as an avocation rather than an all-consuming endeavor. Although I wouldn't describe these next attributes by using pejorative phrases such as "negative side of the ledger", the impact of mathematical quests on both the individual and society is considerably less. Whereas quests to discover the causes and cures for diseases are understandable and attractive to practically everyone, the quest to discover mathematical formulas or proofs is certainly nowhere near as comprehensible or universally attractive.

But you wouldn't be reading this book unless you had at least some interest in it. What I hope to do is not only describe some of the great quests in mathematics but also explain why and how these quests have attracted so many people over the ages. Presidents, emperors, and even Jean-Luc Picard, captain of the starship *Enterprise*, have been among those so captivated by these quests.

How Much Math Do I Need to Read This Book?

This is a book that focuses more on telling what was done – or not yet done – and who did it rather than on how it was done. Arithmetic, simple algebra, and a little geometry will suffice to understand most of the math you see in the book. There are instances where you will see math you may not have come across before, but arithmetic, simple algebra, and a little geometry will suffice to follow it. There is some first-semester calculus that shows up, but if this is unfamiliar to you, feel free to just skip to the conclusion. As a useful parallel, in order to understand how the mRNA COVID-19 vaccines work, you need to understand what messenger RNA is and how it functions; however, if you are just reading a history of the recent pandemic, you can skip over that stuff and still get a sense that this was a great development, resulting from a deeper knowledge of biochemistry.

The Devil and Simon Flagg

Channeling Robert Palmer, I might as well face it: I'm addicted to math. And *The Devil and Simon Flagg* is, IMHO, the best story ever written about a mathematical quest and how it addicts the seeker – at least, the best one I've ever read. Others feel that way, too, as the story has been anthologized and can be found on a number of websites.

It was written by Arthur Porges, who was a prolific writer of science fiction and mystery stories. Almost all of Porges' stories contain brilliant twists. One of his mystery stories is called *The Nose of the Beagle*. It takes place in the late nineteenth century and is about a policeman in a small town in England who is baffled by a murder that has taken place. He learns that there's a really bright guy, a loner, who lives at the edge of town and who might be able to help, and so he goes to him to discuss the case. It isn't until the story has almost ended that the reader realizes that the loner is Charles Darwin. This is typical of the type of stuff that Porges wrote, and I count myself as one of his earliest – and greatest – fans.

During the early 1990s, I conceived of a different way to teach liberal arts mathematics (a.k.a. math for poets), involving having the students read stories about different mathematical topics. I contacted a publisher; they liked the idea and sent me an advance, and I began to work. I wanted to use *Flagg* as an introduction to the book and contacted Porges through my publisher.

It is customary in such situations for an author to grant such a right for a fee, but Art wrote me an absolutely charming letter saying that he would be honored to have his story appear in the book, and I needn't pay anything for the privilege. Sadly, the project was canceled when the publisher was taken over by a larger publisher, but the book – which remained dormant on my hard drive for two decades – finally made its way to print in 2016 as (shameless plug) *L.A. Math: Romance, Crime, and Mathematics in the City of Angels*. But Art and I corresponded for more than a decade, and he was a delightful pen pal. He had actually been a part-time math instructor at both Occidental College and L.A. City College but had moved to Pacific Grove, a small town in the Monterey area known for its attractiveness to monarch butterflies, when he discovered that he could make a living as a writer.

Art would type his letters on an old-fashioned typewriter, which invariably needed a new ribbon; by the end of the letter, the words would often be more physical impressions on the page than ink, but I relished every time a letter from him appeared in my mailbox. I still have the beautiful pen he gave me as a wedding present; it lasted more than 15 years. But I think that the charm of *Flagg* will last as long as there are people to read it.

I originally got in touch with Porges in the early 1990s – at a time when developments were taking place in both the world as a whole and the mathematical world, which made it a good idea to slightly revise the original story. Art agreed, and together we made the revisions. My original intention had been to print the revised story here, but in order to do so I would need to obtain the approval of the executor of Art's literary estate, as Art passed away in 2006. I have written to him, but he has not responded. Every so often emails get lost, but another possibility is that the executor of Art's literary estate has also passed away.

At any rate, I have been unable to obtain that permission, and so instead of reprinting the story, I will summarize the revised version that he and I wrote. However, the summary will lack the charm of the original story, which was written with Art's characteristic warmth and humor. If you choose to read the original story, which can be found in several anthologies and also by browsing the internet, I am sure you will find that the story well repays the effort you make to find it and the time it takes to read it.

The Devil and Simon Flagg is a story about three charming individuals. Simon Flagg is a mathematics professor, and his wife is also a professor at the same school of either literature or history. The third character, of course, is the Devil, and I am quite sure that the Devil was never depicted in literature in quite the way that Porges did. More about that as we continue with the story.

The story begins with Flagg's wife garnering enough knowledge from old medieval tracts to summon the Devil, who appears in Flagg's living room. After some initial negotiation, the Devil agrees to one of the classic "business deals" that the Devil is presumed to have on hand. The Devil agrees to try to answer a question that Flagg poses. If he cannot do so in twenty-four hours, he would guarantee health, wealth, and happiness for Simon and his wife. On the other hand, if he does answer the question, the Devil would gain possession of Simon's soul.

A cute point in the story is that the Devil includes a no-paradox clause in the contract, which is of course signed in blood in order to forestall Simon from asking a question which has no answer, such as the Barber Paradox that arose during the Middle Ages: If the barber shaves every man in the village who does not shave himself, who shaves the barber? If you work through the two possibilities – either the barber shaves himself or he does not – you'll see that each leads to a logical contradiction. Having dispensed with that possibility, the Devil eagerly awaits Simon's question and is stunned when Simon asks him if the Goldbach Conjecture is true.

The Goldbach Conjecture was originally propounded by Christian Goldbach in the seventeenth century. It's really simple: Is every even number the sum of two primes? For small even numbers, such as 20, it's relatively easy to answer the question. For example, 20 is equal to 13 plus 7, and both are primes.

In order to satisfactorily answer the question, the Devil would have to do what any mathematician would have to do to answer it: either construct a mathematical proof of the assertion or find a counterexample – an even number which is not the sum of two primes. We quickly learn that the Devil studied at Cambridge in his youth but seemed to have mostly taken courses in philosophy – and he did really poorly in school at geometry! Simon acquaints the Devil with the problem, and the Devil disappears, presumably to work on it.

Every few hours, the Devil reappears in Simon's living room with progress reports on how he is doing. The first appearance comes with an announcement that he has mastered algebra, geometry, and trigonometry. Every reappearance is greeted by Simon's introducing the Devil to other branches of mathematics. The Devil asks if it is necessary for him to learn all these obscure branches of mathematics, and Simon replies that he has no idea but the great mathematicians have tried them all, and you never know what will work to solve a problem. These reappearances make the Devil not unlike an extremely talented student who says, "Do I have to know that for the test?" – an experience common to every math teacher. And Porges was one, until he decided to become a full-time writer.

Finally, the twenty-four hours are up. An angry and exhausted Devil appears in Simon's living room, holding a sheaf of papers. He hurls them in disgust on the floor and informs Simon that he has won the bet. Not only was the Devil unable to solve the problem, but even mathematicians on more advanced planets have been unable to do so. The Devil disappears, and Simon and his wife embrace.

But that's not the end of the story. Simon pulls away from the embrace and picks up the papers the Devil has thrown on the floor. Simon is a mathematician who has tried himself to solve the Goldbach Conjecture, and he wants to see what approaches the Devil may have tried.

At that point, an embarrassed Devil reappears and asks Simon if he could look at the papers. In the process of trying to win the bet, the Devil has become obsessed with solving the riddle of the Goldbach Conjecture. He tells Simon that he is so close, he only needs to prove one simple lemma to wrap it up. Noticing Simon's interest, he mentions a particularly promising line of attack that he was in the process of trying. They adjourn to a convenient table to pursue this approach, as Mrs. Flagg observes that another all-night session seems to be looming – as so often happens when mathematicians get together to work on a tantalizing problem.

When the Devil reappears for the last time, Porges has him describe how wrapped up he has become in the Goldbach Conjecture by saying, "It certainly gets you". In those four words, Porges summed up the addiction of the quest – whether it be in mathematics or anywhere else. The quest for the solution to the Goldbach Conjecture is one of the ongoing quests I'll describe later in the book. Any quest that has been going on for centuries, as has this one, is certainly worthy of the label "quest" – a long and arduous search.

Truth and Beauty

The great English poet Keats declared, in *Ode on a Grecian Urn*, "Beauty is truth, truth beauty, – that is all / Ye know on earth, and all ye need to know". Certainly poetic, and it may indeed be a beautiful statement, but it fails the truth test and the "all ye need to know" test under the most cursory examination. Beauty, as is well known, is in the eye of the beholder; truth is absolute. And as to "all ye need to know", well, you'd better know a good plumber when the toilet backs up.

But some mathematical truths do indeed possess beauty, not only in what they say but in how they are established. I don't think even the most ardent algebraist would claim that there is something beautiful about the fact that the only solution to $3x + 1 = 7$ is $x = 2$. But it is universally agreed – at least among mathematicians – that there is a profound beauty to the Pythagorean Theorem that, in a right triangle, the square of the hypotenuse is equal to the sum of the squares of the other two sides. Not only is it simple to state, it also tells us something highly useful about our world. Cosmologists even use it to characterize the geometrical nature of our Universe.

Beyond that, there are over 400 proofs of this theorem, and some are exquisitely beautiful in that they use very little in the way of additional geometrical results and assemble the logical structure of the proof. This is the same beauty we experience when we hear Sherlock Holmes declaim, upon seeing Watson return one evening, that Watson does not propose to invest in South African securities. Holmes observes the pieces of the puzzle and puts them together beautifully to arrive at an elegant conclusion – much as the beautiful proofs of the Pythagorean Theorem do.

There is a beauty to the statement of the Goldbach Conjecture, the results that have been proved so far, and how the quest has proceeded. This book is the story of this – and other – mathematical quests. All have been long and arduous searches. Some have ended successfully, some not so. Some have been completed; others are still ongoing. Some are so simply stated that even a reader with only high school mathematics can pursue them if so inclined. But all these quests, like all the quests upon which humans embark, have their own stories of triumph and disaster, and they involve people who are every bit as fascinating as the heroes and villains who populate our dramas and our history.

The Value of Mathematical Quests

It's easy to understand the value of a scientific quest because the payoff is so immediate. When we understand the nature of a disease, we can take steps to cure it. When we understand electricity, we can build devices utilizing its awesome potential. And, of course, that's what we've done throughout our history.

But why should we care about whether or not mathematicians are able to find a particular formula? Let me illustrate with an example which is mostly rooted in reality, but in order to make the potential value of unearthing a formula absolutely clear cut, I've stuck in something that might already have happened without my knowing or might actually still happen.

The story starts with the French mathematician and physicist Jean-Paul Fourier's investigations of the mathematical description of the temperature distribution in a heated object. The equation that describes the temperature distribution in a heated bar is known as the one-dimensional heat equation. It is a partial differential equation, and partial differential equations are notoriously difficult to solve. In fact, the quest for the solutions to one particular very important partial differential equation, the Navier–Stokes equation, is an ongoing quest to this day, which will be described later in this book.

But back to Fourier. Over the years, mathematicians have compiled "databases" of different types of functions, which can be useful in describing various phenomena. The simplest such functions are polynomials, which are well known to everyone who has ever taken any math course higher than basic arithmetic. Another extremely important class of functions is the trigonometric functions, and it was by using the trigonometric functions sine and cosine that Fourier was able to express solutions to the one-dimensional heat equation.

Enter the mathematicians, who now had a new quest: Which functions could be described simply by using sines and cosines? They discovered that periodic functions could be so described, and if you've ever seen a cardiogram, you've seen a periodic function because our hearts generally beat pretty regularly, repeating the same pattern over and over every few seconds.

This pattern is a periodic function and can be described simply by using sines and cosines. And it is at this point that I'm going to simplify the discussion.

Let's assume that the function $H(t)$ describes the amplitude of the heartbeat – how strong it is and where it is in the heartbeat cycle – as a simple representation of the form

$$H(t) = A \sin t + B \cos t.$$

Medical researchers now compile tables of these heartbeat functions, correlate with patient histories, and observe something: Those patients for whom the ratio A/B is less than 0.1 are much more prone to atrial fibrillation.

This greatly increases the value of a patient's cardiogram as an advance warning of a possible dangerous condition.

This is an isolated simple example, but it's the type of investigation going on in any subject that is amenable to mathematical modeling. And because of the universality of mathematics, that means practically everything.

So, some mathematical quests have proved highly valuable. Others, probably not. Then, why not concentrate on those which have potential value and not waste time on stuff like the Goldbach Conjecture, which may be interesting but doesn't seem like it would have much practical importance?

Let's ask arguably the most famous scientist of all time, Albert Einstein, to weigh in on that. Back in the latter part of the nineteenth century, a bunch of Italian mathematicians were attempting to describe surfaces using a branch of mathematics known as differential geometry. At that time, differential geometry was an interesting but relatively minor area of mathematics that didn't seem to be of any particular relevance to important stuff. But it turned out that it supplied precisely the right tools for Einstein to formulate his general theory of relativity. And if you think that relativity isn't important, just imagine removing our knowledge of it; a lot more people would be driving over cliffs and into rivers because their GPS devices would be giving them erroneous information.

This story has been repeated time and again – and even happened to me. I spent the first half of my career studying problems involving continuous linear operators on Banach spaces, and what papers I wrote were generally read only by mathematicians interested in this. I knew many of them either by name or by reputation, as the problems on which I was working were of interest only to a small segment of the mathematical community. But, in the late 1970s, I started receiving reprint requests from electrical engineers all over the world. Somebody had taken the things I discovered on my quest and discovered that they had applications to signal processors. Who knew?

Beyond Truth and Beauty – Or Maybe Ahead of It

I'm talking about fun.

Most of us, at one stage or another, become obsessed with something we enjoy; it may be golf, playing piano, or going fishing. We look forward to the opportunity to do it, not because we expect to become rich or famous doing it but because it's fun.

There are a lot of people who have enjoyed doing math – emperors, presidents, and Captain Jean-Luc Picard, too. Of course, this book focuses on the great quests of mathematics, both completed and ongoing, but it also discusses how the reader can get in on the fun. And what's more, you can do this with almost no training whatsoever.

One of the great discoveries of science occurred when Anton von Leeuwenhoek, who was an ordinary tradesman in seventeenth-century Holland, got hold of a primitive microscope and decided to use it to take a look at a drop of pond water. He discovered something never seen before: the world of protozoa. His name is forever enshrined in the history of science for this discovery.

The microscope was the state of the art in seventeenth-century technology. Not everyone could obtain one. But you have the advantage of access to an extraordinary tool which could enable you to inaugurate quests that will not only be immensely enjoyable but may put your name up there among the famous (although probably not the rich and famous). And when the time comes, I'll tell you what this extraordinary tool is (if you haven't already guessed) and how you can use it to embark on quests of your own.

For mathematicians, the quest for truth and beauty is often an end in itself. But those quests have often paid dividends in the past and certainly will continue to pay off in the future. And for those interested in the Goldbach Conjecture, although at the moment it may seem only of academic interest whether every even number is the sum of two primes, the potential value of this could show up in fields we are currently exploring or those we have yet to discover.

But the one thing almost every mathematician will tell you about any quest upon which they ever embarked is how much fun it was. And that accords perfectly with my philosophy of life: Try to leave the world a little better than you found it and have fun in the process.

I'd like to thank two people in particular for helping to make this book possible. One is my wife Linda, who deserves thanks not only from me but from her students. In the post-pandemic era, many teachers have succumbed to doing quick and dirty teaching by doing so online. Linda still teaches the old-fashioned way, face-to-face with meticulously prepared lectures and a willingness to go out of her way to help her students learn and appreciate mathematics. Both Linda and I would agree that anyone who wants to learn mathematics can do so far better from a book than from an online course – and far better from a real-life teacher than from a book.

The second person is Al Posamentier. I've done podcasts with Al on his books for almost a decade, and last summer I finally got to meet him when we shared a delightful dinner in New York. Al not only urged me to write this book but managed to convince World Scientific to publish it. As my father was fond of saying, Al is a gentleman and a scholar – and unfortunately, there are very few of us left.

About the Cover

Mathematics has been called the science of patterns, so it probably won't come as a surprise that the cover incorporates a number of patterns. Even though some or all of these patterns may be unfamiliar to you, simply observing the diagram and knowing that a pattern was used to construct it should be enough for you to work out what belongs in place of the question mark. But if puzzles don't intrigue you, the answer will be revealed at the end of this section.

The Pythagorean Theorem

Look at the diagram in Column 2 of Row 1, the one in which the interiors of all four triangles are uncolored. The yellow square border around each of the 16 diagrams can be thought of as a frame, and the interior enclosed by the frame as a tile. Each tile consists of the outer square, four congruent triangles, and an inner square. We will label the hypotenuse of a triangle as c, the longer side of the triangle as a, and its shorter side as b.

It is easy to see that the side of the outer square is $(a + b)$, so its area is $(a + b)^2 = a^2 + 2ab + b^2$. But this area is the sum of the areas of the inner square, whose side is c, and the four congruent triangles. This area is $c^2 + 4(\frac{1}{2}ab) = c^2 + 2ab$. So, $a^2 + 2ab + b^2 = c^2 + 2ab$, and subtracting $2ab$ from both sides yields $c^2 = a^2 + b^2$. This is the Pythagorean Theorem, one of the greatest mathematical truths, and the proof given here is considered by many mathematicians to be a quintessential example of a beautiful and elegant proof.

Binary Numbers

Binary numbers use only 0 and 1 to construct a number. The decimal number 123 tells us that it is the sum of one hundred (a hundred is 10^2), two tens (a ten is 10^1), and three ones (a one is 10^0). The binary number 1011 corresponds to one eight (an eight is 2^3), zero fours (a four is 2^2), one two (a two is 2^1), and one one (a one is 2^0). Its decimal value is $8 + 2 + 1 = 11$.

The triangles have been colored using binary numbers: An uncolored triangle represents the binary digit 0, and a colored one represents the binary digit 1. The triangle in the upper-right corner represents the number of eights. Moving clockwise, the triangle in the lower-right corner represents the number of fours, the triangle in the lower-left corner represents the number of twos, and the triangle in the upper-right corner represents the number of ones. The diagram in Column 2 of Row 3 has only the lower-right triangle (which represents the number of fours) uncolored, so the binary number it represents is 1011: one eight, zero fours, one two, and one one, corresponding to a decimal value of $8 + 2 + 1 = 11$.

Magic Squares

A magic square is a square array of numbers in which all rows and columns add up to the same number. The magic square used to construct the cover is

9	0	7	14
15	6	1	8
2	11	12	5
4	13	10	3

Note that the sum of each row and column is 30. The number in Column 2 of Row 3 is 11, so the coloring of the triangles corresponds to the binary number 1011, as discussed in the previous section.

The Missing Piece

Probably the simplest way to solve the puzzle is to realize that there are 16 different ways to color the triangles, from leaving all triangles uncolored (Row 1, Column 2) to coloring all four triangles (Row 2, Column 1). The only missing way to color the triangles is to color only the upper-right and lower-right triangles. Check it out.

Part I

The Quests of Antiquity

Hollywood has programmed us to think that the great quests of antiquity involved such artifacts as the Lost Ark of the Covenant. Stories such as the quest of Jason and the Argonauts for the Golden Fleece are a part of classical mythology, and you just have to look at the first four letters of the word "mythology" to realize that this is just a story.

The latest quest of Indiana Jones, *Indiana Jones and the Dial of Destiny*, involved a gadget called the Antikythera of Archimedes. I'm not sure whether there ever was anything called the Lost Ark of the Covenant, but Archimedes – arguably the greatest mathematician and scientist of the ancient Greeks – really did invent the Antikythera. Spoiler alert: In the movie, the Antikythera reveals and manipulates time fissures, a phenomenon beloved of sci-fi devotees but decidedly lacking in the real Universe. The real Antikythera was an incredibly ingenious analog computer for predicting astronomical positions. In addition, it could also predict when the next Olympic games could be held. It was actually discovered about 30 years ago by an archaeologist and was constructed of wood. Archimedes, in addition to being a brilliant mathematician and scientist, was also a great constructor of gadgets.

If I were to guess, the ancients were much too busy with the difficult business aptly described by the title of the song *Stayin' Alive* to concern themselves with quests for things which did not necessarily exist. But as far as we know, they were the first to describe what is the main subject of this book: the great quests of mathematics.

Mathematics most likely began as the result of a search for answers to simple questions of arithmetic involving addition and multiplication. But knowledge not only closes doors by answering questions; it also opens them by prompting additional questions. When Max Planck, the guy who was responsible for quantum mechanics, initially decided to attend college, he asked Philipp von Jolly, one of the leading physicists of the day, what he felt was the future of physics. Von Jolly's outlook was grim. He felt that physics had answered all the questions there were, and all that remained to do was tack on a few more digits to the right of the decimal point in the values of the various physical constants, such as the speed of light. In less than half a century, X-rays were discovered, opening the door to sub-atomic physics, Einstein came up with the special and general theories of relativity, and Max Planck invented quantum mechanics – and it was off to the races.

Mathematics has never suffered from a paucity of questions. Some of its questions are answered fairly quickly. Others can take centuries – or even millennia. And there were three great questions that the Greek geometers propounded, which indeed took millennia to unravel. To be fair, a large chunk of one of those millennia was occupied by the Dark Ages, during which not a whole lot of people were thinking about mathematics. Even so, it's instructive – and fascinating – to see how the first great quests of mathematics arose and how they were eventually unraveled.

Chapter 1

The Great Geometric Puzzles of Ancient Greece

The COVID-19 pandemic was the greatest challenge mankind has faced – so far – in the twenty-first century. We came through it relatively unscathed – at least when compared with the Spanish flu pandemic of the first part of the twentieth century, which killed over one hundred million people. There were a number of reasons why we did so well. First, the civilized world generally had good public health measures and understood the nature of disease and disease transmission. Second, through the efforts of the worldwide medical and scientific communities, an effective vaccine was developed to prevent or mitigate the effects of COVID-19. Finally, drugs and treatments became available after a couple of years to help people who had contracted the disease.

Things were considerably different two and a half millennia ago, when Greece was visited by the first widely documented pandemic.

Doubling the Cube

The year was 410 BC, and Athens was busy fighting the Peloponnesian War when the pandemic struck. Hippocrates, the father of Greek medicine, was alive at the time, but there was nothing Hippocrates or any other physician could have done to fight typhoid fever, which eventually killed almost one out of every three physicians. Thucydides, perhaps the first great historian, actually contracted the plague himself but survived it and gave us a pretty good description. The eyes, throat, and tongue became red and bloody.

This was followed by sneezing, coughing, diarrhea, and vomiting. In the final stages, the skin was covered in ulcerated sores and pustules, accompanied by a burning, unquenchable thirst – nasty. Nowadays, we have vaccines to prevent it, but if you should contract it and it is diagnosed in the early stages, antibiotics should take care of it. But, like COVID-19, it's best to get the vaccine to prevent getting the disease.

With no medical defenses, the Greeks resorted to consulting the oracle at Delos, who recommended doubling the size of the existing altar, which was in the shape of a cube. Now, if you have a cube and you double the length of the edge of the cube, since the volume of a cube is the cube (not surprising) of the length of the side, the volume will actually increase to eight times the original volume. In order to double the volume of the cube, it was necessary to produce a side whose length was equal to that of the original side multiplied by the cube root of 2.

Ruler-and-Compass Constructions

Anyone who has taken a course in plane geometry will be familiar with ruler-and-compass constructions. Their history dates, not surprisingly, back to the Greek geometers, who were also responsible for laying down the rules for using these two implements. The only non-obvious rule is that you're not allowed to put a mark on the ruler – or whatever you're using as a straightedge. In fact, the Greeks probably would have disallowed modern rulers because they come with marks already on them.

The Greeks never doubled the size of the original altar – or at least they didn't during the plague, which lasted four years. But there was still the problem of multiplying a given length by the cube root of 2, and this problem was soon solved by a number of people using different methods but which did not conform to the rules of ruler-and-compass constructions. Archytas came up with a construction that involved the intersection of three surfaces: a cylinder, a cone, and a torus (which has the shape of a doughnut). Menaechmus, of whom you've probably never heard, and Hippocrates (the father of Greek medicine and the guy responsible for the Hippocratic Oath that all doctors take) managed to find solutions involving intersections of hyperbolas and parabolas. Hyperbolas and parabolas were well known to the

Greeks because they are the intersections of a plane and a cone. Possibly Hippocrates' interest in this problem arose as a result of his interest in medicine. Obviously, the plague was a medical problem, and since the oracle at Delos had suggested the way to cure the plague was to duplicate the cube, this likely stimulated Hippocrates to take a look.

Perhaps the earliest polymath was Eratosthenes, to whom the Greeks gave the nickname Beta because in any area of Greek intellectual achievement, Eratosthenes was the next guy you would talk to if you couldn't get hold of Numero Uno. Eratosthenes not only came up with a relatively simple method to find the cube root of a number using rotations of lines and attached triangles, but his method could also be used to find any other integer root, such as fourth roots or nineteenth roots. Eratosthenes also told us what he thought of his rivals' solution to the problem of finding cube roots:

> *If, good friend, thou mindest to obtain from any small cube a cube the double of it, and duly to change any solid figure into another, this is in thy power; thou canst find the measure of a fold, a pit, or the broad basin of a hollow well, by this method, that is, if thou thus catch between two rulers two means with their extreme ends converging. Do not thou seek to do the difficult business of Archytas's cylinders, or to cut the cone in the triads of Menaechmus.* [1]

At any rate, should the plague return and it prove necessary to construct an altar twice the size of the original, the Greeks could do it. But no one seemed to be able to do it using a standard ruler-and-compass construction.

Regular Polygons

Anyone who has taken a course in geometry will be familiar with the concept of a regular polygon: a polygon with all sides of equal length and all interior angles are equal. The two most familiar regular polygons are the equilateral triangle and the square, but we do see other regular polygons in the real world. Perhaps the most famous is the Pentagon, the giant five-sided office building in Arlington, Virginia, which houses numerous government offices.

---

[1] https://mathshistory.st-andrews.ac.uk/HistTopics/Doubling_the_cube/.

It wasn't designed in the shape of a pentagon because the architects were lovers of geometry but because the land on which it was to be built was bordered by five roads. Another well-known regular polygon is the eight-sided octagon, which, when colored red, is almost universally used as a stop sign – with the exception of Japan, which uses an inverted triangle.

The Greek geometers were able to construct equilateral triangles, squares, and regular pentagons using ruler-and-compass constructions. It isn't clear that the problem of which regular polygons were amenable to ruler-and-compass constructions had assumed the nature of a quest for them; it certainly didn't have the urgency of doubling the cube because the oracle at Delos hadn't suggested constructing regular polygons to cure the plague. As a result, the problem of which regular polygons could be so constructed remained in the "unsolved problem" category until one of the greatest mathematicians in history became intrigued by the problem two millennia later.

Carl Friedrich Gauss

When he was seven years old, Gauss' arithmetic teacher left the room for a short period. In order to keep the children busy, the teacher instructed them to add the numbers from 1 to 100. When he returned a short while later, the teacher noticed that the other students were still busy, but Gauss had written 5,050 – the correct answer – on his slate. The astounded teacher asked how Gauss had managed to add the integers so rapidly. Gauss had noticed that there were 50 pairs – 1 and 100, 2 and 99, 3 and 98, . . . , 50 and 51 – and each pair added up to 101. So, the problem reduced to multiplying 50 by 101, and that's why Gauss was able to come up with the answer so quickly. This observation is known to mathematicians as the "Gauss trick", and while many mathematicians have techniques named after them, Gauss is the only one who entered the record book at the age of seven.

Some time later, Gauss became interested in the problem of constructing regular polygons using a ruler and compass; the polygons which can be so constructed are called "constructible polygons". Mathematical nomenclature is sometimes arcane, but in this instance it makes perfect sense. Gauss devised a technique which enabled him to construct a regular 17-sided polygon. Some time later, he expanded the technique further.

We've come across prime numbers in the Introduction. Primes themselves fall into a number of categories, one of which is the Fermat prime (another example of a mathematical concept being named for someone), which is a prime number of the form $2^{(2^n)} + 1$. When $n = 0$, this expression is equal to 3; when $n = 1$, it is equal to 5; and when $n = 2$, it is equal to 17. Gauss' 17-sided polygon fits nicely into this framework, and some years later Gauss was able to show that he could construct a regular polygon with n sides as long as n was either an even number, a Fermat prime, or a product of even numbers and Fermat primes.

This established what mathematicians call a sufficient condition for an n-sided regular polygon to be constructible. But it wasn't clear that the condition was necessary, or, in other words, that these were the *only* n-sided constructible regular polygons. This quest was completed – as were several of the other quests in this chapter – by another child prodigy.

Pierre Wantzel

Carl Friedrich Gauss was unquestionably one of history's greatest mathematicians, but he is not the star of this particular chapter. That role is filled by Pierre Wantzel, of whom I had never heard until I started writing about mathematics when I was in my 50s. That's partly because, unlike Gauss, Wantzel's contributions – and impressive contributions they were – were limited to a very narrow area of mathematics, which just happens to be the subject of this chapter.

Biographies of Wantzel often include the phrase "at a remarkably young age". He displayed an impressive aptitude for mathematics as a child and entered college before he entered puberty. While only 15 years old, he edited the second edition of a widely used arithmetic textbook, not only performing the standard editorial function of accuracy-checking but also improving a method for finding square roots.

Wantzel exhibited a remarkable aptitude for practically anything academic. He seems to have achieved first place in practically every academic competition he entered, including first place in a French dissertation at the Collège Charlemagne and first prize in an open competition in a Latin dissertation. He was also placed first on the entrance exam to the École

Polytechnique and on the science section at the École Normale, a feat which had never been previously achieved. He also graduated from the College of Engineering but told his friends that he would only be a mediocre engineer and preferred to teach mathematics. In fact, he made a number of notable contributions to the mathematics associated with engineering problems.

But Wantzel first enters this chapter by polishing off what Gauss had started, demonstrating that the only constructible polygons were the ones Gauss described in terms of even numbers and Fermat primes. However, we could have introduced him earlier when we discussed the problem of duplicating the cube. Gauss had stated that this could not be achieved by means of a ruler-and-compass construction, but Wantzel was the individual who actually proved this.

Sadly, Wantzel died at the age of 34, with much of his potential unfulfilled. A number of talented mathematicians have made significant contributions while in their 70s; Wantzel was almost certainly capable of joining their ranks. One of his biographers stated that Wantzel's death was due to overwork and that his mathematical achievements were so limited because he was so remarkably talented in a number of areas that he could not stay focused on mathematics long enough to produce deep results, which generally only come from spending substantial time studying a problem. We shall see several results in this book that were only achieved after at least a decade of concentrated study.

Squaring the Circle

There have been a number of instances where people worked on mathematics while in prison. Possibly the first recorded instance of this was the imprisonment of the Greek pre-Socratic philosopher Anaxagoras. Back then, philosophy covered a multitude of disciplines, and Anaxagoras is probably best known for elucidating the true causes of eclipses. But he was also imprisoned for impiety, and during that period he decided to work on the problem of squaring the circle: constructing a square whose area was equal to the area of a given circle.

My first teaching job was at UCLA, and as the most junior faculty member, I was assigned the task of dealing with the letters (that's how long ago it was) that people wrote to UCLA's Mathematics Department claiming to have solved problems that are known to have no solution. I must admit I wasn't thrilled to be assigned this task, but I consoled myself by thinking that Albert Einstein spent a few years as a patent examiner, and he undoubtedly had to pass judgment on people who claimed to have invented perpetual-motion machines, or the like. Einstein said he enjoyed those years, as it gave him enough free time to work on physics. Einstein and I also had this in common: We were being paid to do this particular job. Here, though, our paths diverged; whereas Einstein produced world-shaking theories, I was only able to produce a few theories that at best generated very modest tremors.

The impossibility of squaring the circle has reached the level of metaphor, but the word "impossible" seems to admit different interpretations. "Impossible" in mathematics means just that – end of story. However, there are some people who interpret it as "the difficult we do today, the impossible takes a little longer". Some of the letters that the UCLA Math Department received claimed to have solved the problem of squaring the circle.

Many of the letters were along the following lines. Suppose you have a circle of radius 1, then its area is $\pi \times 1^2 = \pi$. If a square has side $\sqrt{\pi}$, its area will be π, so the problem reduces to constructing a line segment of length $\sqrt{\pi}$. The following approach was quite popular.

Start with a circle of radius 1. Its circumference is 2π, so if you take the top half of the circumference, it will have a length of π. Straighten it out, and you have a line segment of length π. There are now a number of constructions that will produce a line segment of length $\sqrt{\pi}$. The one I learned in high school goes as follows: Construct a line segment ABC, where the length of AB is 1 and the length of BC is π. Let this be the diameter of a circle. The perpendicular segment from B to the circumference of the circle will have a length of $\sqrt{\pi}$.

The above is a great proof, except for one important detail: that bit about straightening out the top half of the circumference. You could do that if the circle was made out of wire – assuming you had a wire cutter – but how do you do that using only a ruler and compass? If the problem were to square the circle, the proof above would be satisfactory, but it must conform to the requirement that you use only a ruler and compass. It's easy to kick a ball over the goal line in soccer, but it only counts as a goal if it goes into the net.

Pierre Wantzel, the star of this particular chapter, was able to show that it was not possible to square the circle using only ruler-and-compass constructions. Then, about 40 years later, the combined efforts of two German mathematicians, Ferdinand von Lindemann and Karl Weierstrass, managed to show that π was a transcendental number. The term "transcendental" in mathematics does not mean awe-inspiring or anything like it; it's just any number that cannot be a root of a polynomial with integer coefficients. By that time, it was known that one could not construct a line segment whose length was a transcendental multiple of the length of a given line segment. Note that this result rules out squaring the circle, but it still allows for the duplication of the cube; the impossibility of duplicating the cube was established by other means.

Trisecting the Angle

The quest of the Greek geometers which completes this chapter is the problem of trisection of an angle: Given an angle A, can the angle A/3 be created by means of a ruler-and-compass construction? Of course, bisecting an angle is an elementary construction students learn in the first semester of a course in plane geometry.

When I first heard of this problem in the plane geometry course I was taking during my sophomore year in high school, I thought of a possible way to solve this problem. Given an angle, construct an isosceles triangle with this angle at the vertex. Then, trisect the side opposite the vertex angle. Trisecting line segments is an elementary ruler-and-compass construction; indeed, it is possible to divide a line segment into any number of segments of equal length. Anyway, now draw a line from the vertex of the isosceles triangle to one of the two trisection points in the middle of the opposite side. At the time, I thought this would trisect the angle, but there's a really simple example using analytic geometry to show that it doesn't.

To see that this method doesn't work, let's use it on a right angle. Position the two sides of the isosceles triangle on the coordinate plane by letting one side be the line segment from $(0, 0)$ to $(0, 1)$ and the other side be the line segment from (0, 0) to (1, 0). The side opposite the right angle is the line segment from (0, 1) to (1, 0). The points that trisects this line segment are (1/3, 2/3), and (2/3, 1/3). If this construction did trisect the right angle, the angle whose sides are the line segment from (0, 0) to (1, 0) and the line segment from (0, 0) to (2/3, 1/3) would form a 30° angle. But the line segment from (0, 0) to (2/3, 1/3) is on the line whose equation is $y = \frac{1}{2}x$, and we know from analytic geometry (or simple trigonometry) that the line $y = mx$ makes an angle θ with the x-axis, where $\tan \theta = m$. So, the line $y = \frac{1}{2}x$ makes an angle $\tan^{-1}(1/2)$ with the x-axis, and this angle is approximately 26.6°.

It is, however, possible to create an angle using ruler-and-compass constructions that gets arbitrarily close to an angle of any desired size simply by repeatedly bisecting a right angle and then adding up enough of them. For instance, to get very close to a 17° angle, bisect the right angle and then bisect the resulting 45° angle. Do this three more times to get an angle of 5 5/8 degrees. Then add three of these together to get an angle of 16 7/8 degrees.

Hippocrates Strikes Again

I'd spent my entire mathematical career – more than six decades – without once hearing the name "Hippocrates". Of course, I knew him as the father of modern medicine and the man who devised the Hippocratic Oath, but to find out that he was also a mathematician of note was, to say the least, surprising. Of course, a lot of the Greek scientists were actually polymaths; Eratosthenes and Archimedes come quickly to mind, so maybe it isn't so surprising.

We've seen Hippocrates before in connection with the duplication of the cube, and here he is again in angle trisection. He devised an elegant geometrical construction to trisect the angle. It wasn't one that could be implemented with just a ruler and compass, but it could be implemented using a marked straightedge. This provided a practical solution to the problem of trisecting the angle. Additionally, Nicomedes and Apollonius devised other methods of trisecting the angle. Nicomedes invented a curve called the conchoid specifically for that process, whereas Apollonius used hyperbolas, a curve very familiar to the Greeks. Recall that one of the solutions to the problem of duplicating the cube involved hyperbolas. Having found a practical solution to the problem of trisecting the angle, the Greek mathematicians turned their attention elsewhere.

As mentioned earlier, Gauss stated that the trisection of the angle using a ruler and compass was impossible, but he did not supply a proof. That proof was supplied by Wantzel, the star of this chapter. Interestingly, some of the ideas that Wantzel used to complete this geometric quest were also used in the completion of the quest in the following chapter, which comes from the realm of algebra. And that's one of the fascinating aspects of mathematics: The different subject areas, such as geometry and algebra, are woven together in a fascinating tapestry. Quests which start out in one subject area sometimes end up in others, to the benefit of both.

Part II

Quests of the Middle Ages and the Renaissance

The flame that was mathematics flickered and almost died – at least in Europe – during the Dark Ages. Scientific questions such as "Why is a rose red?" were answered theologically: They were red to reflect the blood of Christ. And mathematics, which went hand in hand with the development of science for the Greeks, was essentially reduced to practical problems, such as keeping accounts.

Nonetheless, developments did take place. The great Arab mathematician Muhammad al-Khwarizmi is known as the father of algebra. Like the Greeks, al-Khwarizmi was a scientist as well as a mathematician. His keen interest in astronomy did not prevent him from publishing an eminently practical book, *Kitab al-Jabr*, from whose title arose the word "algebra". The focus of the book was on problems of how to distribute assets, such as land and inheritance. Al-Khwarizmi also unwittingly instigated the first great quest of the Middle Ages by developing the quadratic formula – the general solution to the equation $ax^2 + bx + c = 0$.

It took some time for Europe to emerge from the Dark Ages and for European mathematicians to appreciate the subject that al-Khwarizmi had outlined in his seminal work. But with the coming of the fifteenth century, mathematicians began to wonder whether algebraic equations of degree 3 – and higher – had formulas similar to the quadratic formula with which to express the solutions. This was the first great quest of the medieval mathematicians, and it was not to be resolved – and in truly surprising fashion – for several centuries.

The Renaissance was not only an awakening in the arts and the sciences; it was also an awakening in mathematics. The development of calculus by Isaac Newton and Gottfried Wilhelm Leibniz ushered in a totally new direction for mathematics, although it had been anticipated to some extent by the Greeks. Mathematicians also began thinking about properties of mathematical objects and structures, from which arose two of the most famous mathematical quests: the search for a proof to Fermat's Last Theorem and whether or not the Goldbach Conjecture was true.

Mathematics, which had shut its doors for almost a thousand years, was back in business.

Chapter 2

The Roots of Polynomials

Let's start this section with a quest that has all the classic ingredients. Long and arduous – check; it lasted for several millennia. An epic goal – check; the search for a formula which would give the roots for any polynomial. Now, this may not strike you as an epic goal, but there are countless real-world problems which are described by polynomials, and a formula giving the solutions to these equations would save everyone a lot of time and effort.

But what lifts this into a category that possibly no other mathematical quest has ever achieved are the individuals involved and the elements of human drama. This quest features some of the most unique characters ever to populate the mathematical landscape, as well as murders, poisonings, duels, conspiracies, love, and death – the whole nine yards. Couple that with a twist ending that no one expected, and you have a quest that probably has no equal in the history of mathematics – and few quests anywhere.

The Adventure Begins

The first formal theorem of geometry – that vertical angles are equal – was proved by the Greek polymath Thales of Miletus approximately six centuries before the birth of Christ. And that's when I thought that mathematics began – so Eurocentric and so wrong. Nearly two thousand years earlier, the Babylonians had actually begun examining linear, quadratic, and cubic equations, although not in the formal sense of categorizing them, as we do today. Certain practical problems arose for both the Babylonians and the Egyptians, which resulted in the need to find the roots of what we nowadays call polynomials, and it's amazing to see how well they did. The Babylonians, for instance, would solve quadratic equations by completing the square, just as is shown in schools to derive the quadratic formula. As far as we know, though, the quadratic formula was first written down in the ninth century by the Arab mathematician Al-Khwarizmi.

It's not clear when the search for roots of polynomials with integer coefficients, i.e. expressions of the form $a_n x^n + a_{n-1} x^{n-1} + \cdots + a_0$, where all the a_k are integers, achieved the status of a quest, but it was known early on that the search unearthed different types of numbers. The simple linear polynomial $2x - 6$ has the positive integer 3 as a root, but $2x - 5$ requires fractions and $2x + 6$ needs negative integers. And when we move up to quadratic equations – polynomials of degree 2 – we need both square roots and complex numbers. Even before mathematicians started to tackle the general cubic, i.e. the roots of $ax^3 + bx^2 + cx + d$, it was apparent that the roots of $x^3 - 2$, whose solution involves the number needed to double the cube, would include cube roots. And just as quadratics had introduced complex numbers, would new surprises be uncovered by discovering the roots of cubics?

Enter the Italians

It's not clear how much was accomplished in the way of solving cubics between Al-Khwarizmi's enunciation of the quadratic formula and the middle of the fifteenth century. This was certainly due in part to the fact that this period encompasses much of the Dark Ages, when there wasn't a whole lot of math and science going on in Europe or, as far as we can tell, anywhere else.

The first mathematician to make serious inroads into the problem of solving the general cubic was Scipione del Ferro, who was a lecturer in arithmetic and geometry at the University of Bologna. He left no published notes, which was probably due to the fact that practically every aspect of Italian culture at the time was enmeshed in competition and conspiracies. You just have to read Machiavelli to get some idea of how one went about getting ahead in court circles, and climbing the ladder in academia had many similarities. You didn't obtain positions or advancement by collaborating with your peers; you did so by bumping them off via intellectual duels. It was customary for aspiring mathematicians to challenge established ones to an intellectual duel. Each mathematician would propose a set of problems for the other to solve. To the victor, then as now, belonged the spoils.

Del Ferro's great advance was coming up with a solution to what are known as "depressed cubics": those which had the form $ax^3 + cx + d$. Fearing a challenge from an aspiring mathematician, del Ferro kept his powder dry by not publishing his results, but thankfully he did record them in a notebook. Had he been challenged, he would have presented his opponent with a list of depressed cubics. As far as we can tell, though, while in del Ferro's possession, the powder remained dry, as he was never challenged.

Del Ferro would, however, discuss his results with his students. He may have realized that he had something of great importance to pass on to the world, for on his deathbed he revealed the solution to one of his students, Antonio Fior. Fior was nowhere near as good a mathematician as del Ferro, but he did realize that the possession of the solution to the general cubic might enable him to ascend to a valued position in the world of Italian academia. And so he issued a challenge to Niccolò Fontana, who is better known to us as Tartaglia, the "Stammerer". Tartaglia's stammer was the result of a childhood injury inflicted by a sword stroke of a soldier. This affected his speech, but not his mind.

Tartaglia at the time was a well-known scholar and possibly accepted Fior's challenge on the theory that it wouldn't be much of a contest and dispensing of this rival would dissuade others from challenging him. So, he constructed a list of 30 assorted questions for Fior – and undoubtedly received the shock of his life when he looked at the questions Fior had submitted to him. Fior's list of questions consisted of 30 depressed cubics.

Ever since the Trojans showed up with that horse – and possibly even before – contests have been decided by secret weapons. Tartaglia was faced with just 30 days to find a solution to Fior's secret weapon – the solution to the depressed cubic – that had been given to him by del Ferro. Nowadays, the solution to a depressed cubic is presented as a purely algebraic exercise, but Tartaglia managed to find a brilliant approach to the problem through geometry to cut through this particular Gordian knot. The wager on the outcome of the contest had been 30 feasts to be given by the loser, but in a spirit of generosity, Tartaglia canceled the obligation. As might be expected, Fior faded into obscurity, while the story of Tartaglia's triumph was headline news, or at least what passed for headline news in sixteenth-century Italian academia.

The story of how an algebraic quest was resolved through a geometric approach would repeat itself more than four centuries later. Stay tuned.

Girolamo Cardano

When you think of the Italian polymaths of the Renaissance, the first name that comes to mind is Leonardo da Vinci. Girolamo Cardano is nowhere near as well known, yet not only was he a polymath, as he made major contributions to science and mathematics, he was also one of the most controversial and colorful characters ever to dot the mathematical landscape.

It's worth spending a little time on his biography outside of his contribution to the quest for the solution by radicals, and his contribution to this quest was enormous. He is responsible for the first mathematical approach to the theory of probability and made notable contributions to hydrodynamics, mechanics, and geology. He was a successful lawyer, the personal physician to celebrities of the times, and at one time probably would have won the Albert Einstein award – had such an award existed – for the world's most recognized scientist. He wrote two encyclopedias that were state of the art.

And that's in addition to his personal life, which, if Cardano were better known today, would doubtless make for a pretty entertaining movie or TV show. But, since this chapter is about the quest for the solution by radicals, let's take a look at how Cardano enters the picture.

The news of Tartaglia's victory and solution to the depressed cubic traveled rapidly (for that era) through Italian academia, of which Cardano had already become a part. He was already interested in algebra and managed to persuade Tartaglia to leave his home in Brescia and visit Cardano at his home in Milan. There he pressed Tartaglia to reveal the secret of the solution to the depressed cubic. Repeated entreaties eventually wore Tartaglia down, and so he told Cardano the solution, on condition that he never publish it and encrypt his work on the subject so that no one else could read them. And, yes, they had encryption back then, and Cardano was something of an expert on the subject.

Cardano was also something of a mystic. In his later life, he was thrown in jail for the heresy of casting the horoscope of Jesus Christ. One day, he had a dream that someone important was about to enter his life. Just by chance, a young man named Lodovico Ferrari came looking for a place in Cardano's household – and, then as now, there's nothing like showing up at the right place at the right time. Not only did Ferrari land a position, it turned out that he had a totally unexpected gift: an ability for mathematics. Cardano had been a university instructor and was an excellent teacher, although he was somewhat less than desirable as a colleague, as he had a habit of criticizing and making fun of others.

After Ferrari had learned enough mathematics, he and Cardano decided to tackle the problem of solving the general cubic. The obvious jumping-off point was Tartaglia's solution to the depressed cubic, which Cardano revealed to Ferrari. But talking about math with someone in your household doesn't constitute publicizing it. At least, we can be pretty sure that's how Cardano saw it because he and Ferrari certainly needed the solution to the depressed cubic.

One of the reasons math and science make progress is that it's possible, to paraphrase the immortal Isaac Newton, to stand on the achievements of giants. It's possible to use knowledge already gained to acquire new knowledge. In mathematics, one of the most important tools is the ability to transform an unsolved problem into a solved one.

Good examples of this can be seen in plane geometry. Knowing the area of a triangle enables one to find the area of a polygon by cutting it into triangles. Knowing the Pythagorean Theorem for the length of the hypotenuse of a right triangle can be used to develop the law of cosines, which enables one to find the length of the missing side of a triangle, knowing the lengths of the other two sides and the angle between them.

This technique is valuable, but it's not always possible. It's not possible to transform the problem of solving a quadratic into a problem of solving linear equations; you need something new. It is, however, possible to transform the problem of finding a solution to the general quadratic into finding the solution of a simple quadratic.

Let's see how this works. It's clear that the roots of $Ax^2 + C$ are simply $\pm\sqrt{-\frac{C}{A}}$, with of course the proviso that we may need imaginary numbers if both A and C have the same sign. The general quadratic has the form $ax^2 + bx + c$. Let's say we get lucky, and after a few unsuccessful attempts, we try the transformation (you can think of it as a substitution) $x = u - \frac{b}{2a}$. Watch what happens:

$$
\begin{aligned}
ax^2 + bx + c &= a\left(u - \frac{b}{2a}\right)^2 + b\left(u - \frac{b}{2a}\right) + c \\
&= a\left(u^2 - \frac{b}{a}u + \frac{b^2}{4a^2}\right) + b\left(u - \frac{b}{2a}\right) + c \\
&= au^2 - bu + \frac{b^2}{4a} + bu - \frac{b^2}{2a} + c \\
&= au^2 + \frac{b^2}{4a} - \frac{b^2}{2a} + c.
\end{aligned}
$$

This is of the form $Au^2 + C$, and we know how to find its roots! Having done that, when we substitute $u = x + \frac{b}{2a}$ and simplify, we get the quadratic formula. Textbooks generally prefer obtaining this formula through the technique of completing the square, but doing it this way not only gives you an idea of the value of transformations but also provides a window into the thinking of some of the great mathematicians who engaged in the quest for solution by radicals.

And it's worth noting that we could think of $Ax^2 + C$ as a depressed quadratic – and maybe, if you look at some books written in the fifteenth and sixteenth centuries, they may have called it that. And fortunately for Cardano and Ferrari, they found a way to transform a general cubic into a depressed cubic. Knowing how to solve the depressed cubic, coupled with a little elementary algebra involving the transformation, enabled them to solve the general cubic.

But the hits just kept on coming! Ferrari came up with a brilliant transformation that converted the problem of solving the general quartic – the polynomial of fourth degree – into a cubic. This would be front page news in every corner of the world of mathematics, except for one problem: the oath Cardano had sworn to Tartaglia to keep secret the solution to the depressed cubic.

What Cardano needed now was a way to rationalize using the solution to the depressed cubic without breaking his oath to Tartaglia. Perhaps it helps to have a legal background for situations like this, as Cardano managed to convince himself and Ferrari that since it was del Ferro who had first found the solution to the depressed cubic, he and Ferrari could use it without breaking his oath to Tartaglia. And so, Cardano wrote up the solution in his classic work *Ars Magna* – the "great art". Cardano wrote that although it took him five years to write it, it could well last for thousands.

Cardano took scrupulous care in *Ars Magna* to give full credit to del Ferro and Tartaglia for their work on the depressed cubic, but Tartaglia was enraged that Cardano had broken a sacred oath. Cardano did not respond to Tartaglia's accusations, but Ferrari was enraged and challenged Tartaglia to a duel. The duel was contested on Ferrari's home court, where he had the home field advantage of a friendly crowd, which Tartaglia blamed for his defeat. It happens today in sports contests, but I'm still charmed by the fact that intellectual duels were the World Cup and Super Bowl for the people of sixteenth-century Italy.

Most mathematicians, like most people, live relatively ordinary lives without a whole lot of drama. But there are exceptions – Cardano was certainly one. Cardano's wife, whom he pursued as the result of having a dream about a woman in a white dress, died young. His eldest son was executed for murder, and his younger son was thrown in jail for criminal activities, where he was tortured. OK, so there were some bad things to say about sixteenth-century Italy. Cardano was, as mentioned earlier, jailed for heresy but later received a pardon.

And I don't know of any other mathematicians besides Ferrari who died as the result of being poisoned by their sister.

Paolo Ruffini and Niels Henrik Abel

Some years ago, I came upon the name Paolo Ruffini, and my vision was slightly blurred at the time, so I saw it as Pollo Ruffini. I don't recall, but maybe I had just dined at an Italian restaurant. It then occurred to me that practically any Italian surname makes a great descriptor for a dish in Italian cuisine – Zuppe del Ferro, Scampi Tartaglia, Vitello Cardano, Ravioli Ferrari!

Yes, the above paragraph certainly constitutes a pause in the quest for the solution by radicals, but after the work of Cardano and Ferrari, there was also a pause in that quest by the mathematical community, lasting nearly two centuries. The next logical step was to do for the quintic what Ferrari had done for the quartic: find a transformation which would reduce a quintic to a quartic and then apply Ferrari's quartic formula. But those transformations had become increasingly more complicated, and thus harder to find, if they even existed. Besides, there were competing developments in the world of mathematics. Newton and Leibniz had started to develop calculus, the mathematics of changing quantities. It was evident that calculus had the power to transform the study of the Universe in a way that finding a solution by radicals to polynomials of ever-higher degrees did not. The exploration of calculus had begun, and it's always easier – and more fun – to get in on an exciting development early in the game when it hasn't been thoroughly worked over.

Nonetheless, some of the greats of the era took a shot at the problem of solution by radicals – or at least thought about it. In particular, the great French mathematician Legendre wrote a treatise on solving algebraic equations, in which he said that he intended to get back to the quintic. But it was Ruffini who was the first to realize that solving the quintic by radicals may have been a bridge too far.

Ruffini actually published a paper in which he stated the result – that it was impossible to find the roots of polynomials of degree five or higher using only radicals. However, there was an error in his proof, and unfortunately Ruffini was never notified of this. Had he been so notified, he would have had a chance to fix the proof, and we might be placing his name among the greats rather than the nearly greats.

Nonetheless, Ruffini's paper contained the key insight that eventually led to the resolution of the quest: The important thing was to examine what happened to equations when the roots of a polynomial were permuted. Niels Henrik Abel, an immensely talented Norwegian mathematician, was the first to come up with a proof that the quintic could not be solved by radicals. His proof followed the same lines as Ruffini's but without the errors.

Abel is one of the hard-luck stories of mathematics. Norway not being a hotbed of mathematical research, he decided to pursue his fortunes by going to Paris, where there was a lot of competition. There are some professions which are in constant demand, such as surgeons, and there are some in which the supply always exceeds the demand, such as entertainers. But the job market for mathematicians is highly variable – sometimes there are more jobs than mathematicians, and sometimes it's the other way around. I was luckier than Abel, for I arrived on the scene during the Cold War, when mathematicians, scientists, and computer programmers were all in short supply.

Abel became discouraged in Paris and returned to Norway, where he tragically died of tuberculosis at just 27 years old. Even sadder, he never knew that the papers he had written in Paris had been very favorably received, and two days after his death, his family received a letter notifying him that he was the recipient of an academic position in Berlin.

But it is not just for his proof of the insolubility of the quintic by radicals that Abel is known. He was one of the first to realize the importance of the concept of group. A group is a mathematical structure with only one operation, such as the set of all integers (positive, negative, and zero) with only addition as the allowed operation. As we all know, addition of integers is commutative ($x + y = y + x$), and commutative groups are known as Abelian groups.

A Simple Example of a Non-commutative Permutation Group

Take three cards from a deck of cards: an ace, a king, and a queen. Put the queen at the bottom, the king in the middle, and the ace at the top.

We're going to perform two operations in different orders. The first operation, which we'll call "SWITCH", switches the top and middle cards. The second operation, which we'll call "MOVE", simply takes the top card and moves it to the bottom.

Let's say we first SWITCH and then MOVE. Remember, we're starting with the ace as the top card, the king in the middle, and the queen at the bottom. After SWITCH, the king is the top card, the ace is in the middle, and the queen is at the bottom. We then perform MOVE. After this, the ace is the top card, the queen is in the middle, and the king is at the bottom.

Now, suppose we first MOVE and then SWITCH. As before, we start with the ace as the top card, the king in the middle, and the queen at the bottom. After MOVE, the king is the top card, the queen is in the middle, and the ace is at the bottom. We then SWITCH. After this, the queen is the top card, the king is in the middle, and the ace is at the bottom.

You may remember that, at the start of the book, I mentioned you really don't need much more than arithmetic, simple algebra, and geometry to understand most of the math in this book. You didn't even need any of that to understand the above explanation.

All the different ways that you can rearrange those three cards comprise what is known as the permutation group on three objects. It is non-commutative because two rearrangements – SWITCH and MOVE – done in different orders give different results. The permutation group on three objects is non-Abelian.

Évariste Galois

If we can only attach one name to the final resolution of this particular quest, the name most mathematicians would select would be Évariste Galois. Although Abel had shown that the quintic was not solvable by radicals, Galois developed the underlying theory (now known as Galois theory) and showed that the resolution of the problem depended on the structure of certain permutation groups, which are not surprisingly known as Galois groups.

In the early nineteenth century, it was customary for students receiving degrees to have one-on-one examinations. His mathematics examiner described him as intelligent, possessing a remarkable spirit of research. His literature examiner was stunned at learning this, declaring that Galois knew absolutely nothing and that he believed that Galois had but little intelligence.

Like Ferrari before him, Galois was both a hothead and a romantic. Both were to have catastrophic effects. His hotheadedness with regard to the revolutionary fervor that was sweeping France landed him in jail, and he eventually fell in love with Stéphanie-Felice du Motel, the daughter of a physician. Her name appears in several of Galois' manuscripts.

Galois was challenged to a duel, probably over Stéphanie, was fatally wounded, and died the next day. Legend has it that he spent the night before the duel working on his mathematical notes, but this may well be apocryphal. His brother and a friend copied those notes and tried to get them published by sending them to two of the leading mathematicians of the era, Carl Friedrich Gauss and Carl Jacobi. There is no evidence that they even bothered to look at them. But 11 years after his death, these notes made their way to Joseph Liouville, who announced to the mathematical world that Galois had given a solution, "…as correct as it is deep of this lovely problem: Given an irreducible equation of prime degree, decide whether or not it is soluble by radicals".

The four-thousand-year quest had ended.

Have We Reached a Dead End?

Polynomials are an important class of functions, but so are trigonometric functions and exponential and logarithmic functions. These are the classes of functions one encounters prior to studying calculus. Calculus presents a new and different class of equations – differential equations – describing how quantities change with respect to changes in the variables that define them. That makes differential equations an extremely valuable tool in the sciences and engineering, and for important equations, people want to know their solutions.

The hypergeometric series (this name was bestowed upon them sometime in the seventeenth century) constitute an extremely large class of functions, which are solutions to very important classes of differential equations. In 1872, the great German mathematician Felix Klein showed that the roots of any quintic could be expressed in terms of hypergeometric functions.

Although the particular result may have been somewhat surprising, the general idea – that there exists another class of functions which can be used to express the roots of the quintic – probably wasn't. There are so many different important classes of functions that those roots almost certainly had to be somewhere. And later research discovered that the roots of even higher-degree polynomials lie in even more obscure classes of functions.

So, I'm guessing that not a lot of work is currently going on in discovering which obscure classes of functions contain the roots of the 11th-degree polynomials, or whatever. If indeed someone did discover this, it would probably just fall under the heading of an isolated, not-especially-interesting factoid. But there always remains the possibility that there is some grander unifying principle that would enable one to say something on the order of "the roots of all 47th-degree polynomials can be found by using blah-blah-blah functions". If I had to guess – and I'm a complete non-expert in this area – either there is no such result or if there is, it's too deep a result for us to discover.

But you never know. And possibly centuries from now, that question might generate another mathematical quest.

Chapter 3

The Goldbach Conjecture

We've already encountered the Goldbach Conjecture – or at least the modern version of it – in the Introduction, when Simon Flagg told the Devil that it was the conjecture that every even number is the sum of two primes. But that isn't the way the conjecture was originally formulated, although it's close.

In 1742, Christian Goldbach wrote to Leonhard Euler, proposing that every integer could be written as the sum of three primes. Back then, the number 1 was regarded as a prime, but other than that, they had the same definition of a prime that we do: an integer other than 1 was a prime if its only integer divisors were 1 and the number itself. That guaranteed that every prime other than 1 or 2 would be odd, and so Euler pointed out to Goldbach that a consequence of this remark would be that every even number greater than 2 would be the sum of two primes. In fact, Euler stated in his reply to Goldbach that he regarded this as a completely certain theorem, although he could not prove it.

And that's the form in which we know the Goldbach Conjecture today.

Why Don't We Consider 1 to Be a Prime?

The simple answer is that making 1 a prime messes up the statement of one of the most important theorems in mathematics: the fundamental theorem of arithmetic. You've known this since elementary school; it's the result that every number can be written in one and only way as a product of prime number. It's understood that the "one and only one way" refers to which primes appear in the factorization, not the order in which they appear. We consider $2 \times 2 \times 3 \times 5$ and $5 \times 2 \times 3 \times 2$ to be the same factorization of 60. There is a standard way to write prime factorizations: Use exponents and list the primes in increasing order, so $60 = 2^2 \times 3 \times 5$.

If we were to consider 1 as a prime number, we could write any number of 1s, and thus the uniqueness of the prime factorization would be lost. This wouldn't be a serious loss, but it would mean that many proofs would have to include a bunch of unnecessary arguments to take care of all the 1s that may or may not appear.

Why Are Primes So Beloved by Mathematicians?

There are lots of other interesting types of numbers, and we'll discuss some of them when it's more appropriate to do so. But primes occupy a special place precisely because of the fundamental theorem of arithmetic. They are the basic building blocks for all integers via the process of multiplication.

Every science is interested in discovering the fundamental building blocks. All living creatures are assemblages of cells, so cells are the fundamental building blocks of biology. Every chemical compound is an assemblage of elements, and what the fundamental theorem of arithmetic is to mathematics, the periodic table is to chemistry. We're still working on the building blocks in physics; in the nineteenth century, they were thought to be atoms. In the first half of the twentieth century, we discovered that atoms were assemblages of electrons, neutrons, and protons, while in the second half, we discovered that the neutrons and protons were assemblages of quarks, and the jury is still out on whether we've gone as far as we can go. The questions of dark matter and dark energy still await resolution.

But by definition, the primes are the building blocks of multiplication, and the fundamental theorem of arithmetic tells us that, using these building blocks, there's only one way to build an integer via multiplication.

As we've gone deeper, though, we've discovered other wonderful and useful properties of primes. In fact, your security and, to some extent, your happiness in the twenty-first century depend on prime numbers.

A Mathematician's Apology

No, the apology isn't mine. That heading is in italics because it's the title of a book written by the great British mathematician G. H. Hardy, who lived in the first half of the twentieth century. Hardy is responsible for the discovery of Srinivasa Ramanujan, the great Indian mathematician whose existence had first come to light because of a letter he wrote to Hardy containing some mathematical results in number theory. Hardy described reading that letter as the only romantic moment in his life, saying that some of the results that Ramanujan described were facts Hardy knew to be true, some were results he suspected were true, and some were results that were too beautiful not to be true. Hardy invited Ramanujan to come to England and work with him. Ramanujan's story was made into the 2015 movie, *The Man Who Knew Infinity*.

Toward the end of his life, Hardy wrote *A Mathematician's Apology*, in which he described his life as a quest for truth and beauty. Hardy had spent his professional life working in number theory, a branch of mathematics that Hardy believed to be useless from a practical sense. Hardy made the parallel between his life and the life of an artist who spent his or her life creating and studying things of beauty. That's what Hardy had done, but the things of beauty he studied were numbers, and the things he had created were theorems concerning the behavior of numbers.

It is inconceivable that someone who spent their life studying number theory would not have been interested in prime numbers – and of course, Hardy was extremely interested in them. Some of Hardy's results concern the difficulty of factoring a number that is the product of two large prime numbers. As we shall see, determining whether a large number is prime is difficult, and even when you know it isn't prime, determining its factors can be even more difficult.

And that's what keeps your bank account and your passwords safe.

The RSA Algorithm

Hardy died in 1947, almost certainly still thinking that his life had been a quest for truth and beauty, but with no practical application – at least, none that he could see. But, with the dawn of the electronic computer age, more and more information was being stored electronically, and there was a need to keep that information secure. When there is only one copy of sensitive information and it's stored in a safe deposit box, you can be pretty certain that it's safe. When there are lots of copies of sensitive information and they're available on lots of computers, there's considerable cause for concern, as well as a need for a way to keep this information secure. And 30 years after Hardy died, three mathematicians – Ron Rivest, Adi Shamir, and Leonard Adleman – were able to use the difficulty of factoring large numbers that were the product of two primes to create a highly secure method of encryption and decryption.

The details of the method are not really difficult, but they are a little technical and can be found with a little web browsing. What matters is that it is the difficulty of factoring – the problem that Hardy and other number theorists were investigating – that keeps your passwords and financial accounts secure. Several years ago, a large-scale project was undertaken to see how long it would take to factor a large number of the types upon which the RSA algorithm was constructed. Hundreds of computers were involved – and it took over nine months.

However, there are developments taking place that may make it necessary to find even more sophisticated methods of encryption than the RSA algorithm. You've probably heard of something called quantum computing. Quantum computing, which is not yet a reality on any practical level, is much faster than even the fastest computers that currently exist. But it's coming, and when it does, the RSA algorithm may be vulnerable to attack.

But just like the Y2K scare – remember that? – forewarned is forearmed. Even though classical methods such as the RSA algorithm may be vulnerable to attack through quantum computation, there are undoubtedly quantum algorithms which are more resistant to such attacks.

How Can We Tell If a Number Is a Prime?

There's an easy way that would probably occur to any child who has learned about division. If a number N is not prime, it has a whole number divisor other than 1 or N which won't be larger than $N/2$. So, simply try to divide N by 2, 3, 4, ... up to $N/2$. If none of these numbers divide N, N must be prime.

We can improve on this somewhat by using the fundamental theorem of arithmetic. If a number N is not prime, it has a prime divisor less than or equal to the square root of N. So, assuming we have a sufficiently complete table of prime numbers, we can test N for primality by dividing it by every prime number less than the square root of N.

Alternatively, we may not have a sufficiently complete table of prime numbers. Even so, if there are enough of them, we can test N for primality by dividing them by all the prime numbers in our admittedly incomplete table of prime numbers. Although we couldn't be absolutely sure that N was prime by doing this, the probability that N is not prime might be sufficiently small that we'd be willing to take the chance. After all, this is what statistics is all about. We can't test everyone to see whether a particular vaccine against COVID-19 is effective, but if it is effective against a sufficiently large sample, we can have a great deal of confidence that the vaccine will be effective for the entire population.

There are two ways we can test whether a number is prime. One is known as deterministic because you come up with a definite answer, just as you would if you tried to divide the number by every prime less than its square root. The other is probabilistic because you come up with a probability that the number is prime; it may not offer a guarantee, but like the COVID-19 vaccine, you might judge that the risk is sufficiently small.

But let's go back to where we had a complete table of all the primes less than the square root of the number we were checking for primality. Do we have any idea of how large that table would be?

The Prime Number Theorem and the Frequency of Primes

Euclid showed long ago that there were an infinity of primes through a really simple proof. If there were only finitely many primes, you could list them all. Then multiply them all together and add 1. The resulting number has a remainder of 1 when divided by any prime in your list, and so it must be prime. It's not in the list of primes because it's larger than any of them.

This contradicts the assumption that you listed all the primes, and so there must be an infinite number of primes.

There's an intuitive sense in which some infinite subsets are larger than others. We can describe this by the observation that half of the integers are divisible by 2, but only one-third of the integers are divisible by 3, and only one-tenth of the integers are divisible by 10. Even though we may not have quantified this to the satisfaction of mathematicians, we know what we mean.

There are tables of primes that can be found on the internet. There are 25 primes less than 100, 168 primes less than 1,000, 1,229 primes less than 10,000, and 9,592 primes less than 100,000. So, as you go further out, it gets harder and harder to find prime numbers.

Mathematicians noticed this early on and wondered if there was some way to quantify the number of primes less than a given number. The prime number theorem, which was first proved late in the nineteenth century, can be stated in a number of ways. One way to state this result is that if $\ln N$ is the natural logarithm of N, then the number of primes less than N is approximately $N/\ln N$. Another way is that the nth prime number is approximately $n \ln n$. There is a rigorous definition of "approximately", but for our purposes, the basic idea is good enough.

The idea of density can be defined intuitively without the need for the precise rigor that must appear when it is used in proofs. Suppose that we have an infinite set of integers A. For each N, we can look at the fraction $n(A, N)/N$, where $n(A, N)$ is the number of integers belonging to A that are less than N. The density of A is the limit as N approaches infinity of $n(A, N)/N$.

This definition accords with what we would expect from obvious situations. The density of the even numbers is 1/2, and the density of the numbers divisible by 3 is 1/3, and so on. The density of primes is approximately $(N/\ln N)/N = 1/\ln N$, which approaches 0 as N approaches infinity.

However, there are sets for which this density cannot be computed because the limit does not exist. Here's a simple example of such a set. Let A consist of all the integers from 1 to 3, from 10 to 27, from 82 to 243, ..., from $3^k + 1$ to 3^{k+1}, and so on, where k is even. If we look at the ratio $n(A, 3^k)/3^k$, it swings back and forth from a number $\geq 2/3$ if k is odd to a number $\leq 1/3$ if k is even.

The prime number theorem has an impact on the Goldbach Conjecture. It's relatively easy to find two primes that add up to small even numbers; often, this can be done in several different ways. For instance, $50 = 47+3 = 43+7 = 37+13 = 31+19$. But remember that half the integers are even, and as they get larger and larger, a smaller fraction of the numbers less than an even number are prime, so it seems as if it would be harder to find two prime numbers adding up to a given even number.

And it's hard to find primes – at least, large ones. There is no simple function using positive integer inputs which outputs only primes, although there are really complicated ones. For instance, in 1976, four mathematicians presented an example of a polynomial of degree 25 in 26 variables which does just that. However, the output doesn't give every prime, and the only known functions which do that are essentially equivalent to the function $f(n) =$ the nth prime number. This function is obviously useless for computational purposes.

We now have enough in the way of descriptions to understand the history of the Goldbach Conjecture.

The Early Years

It doesn't seem that there was a whole lot of interest in the Goldbach Conjecture after the initial exchange between Euler and Goldbach. During the later part of the eighteenth and early part of the nineteenth century, two great mathematicians obtained results on how integers could be represented as sums. Joseph-Louis Lagrange showed that every positive integer was the sum of four squares. Carl Friedrich Gauss showed, in his famous "Eureka" theorem, that every integer was the sum of three triangular numbers (triangular numbers are $1, 3, 6, 10, \ldots$ that come from arranging dots in a right triangle, with one in the top row, two in the second row, three in the third row, etc.). But nobody seemed to be taking a hard look at the Goldbach Conjecture until Lev Schnirelmann showed in 1930 that every positive integer greater than 1 could be written as the sum of at most 800,000 primes.

OK, it's a long way from 800,000 down to the 2 of the Goldbach Conjecture. But Schnirelmann's result opened a door, and Ivan Vinogradov, a brilliant Russian mathematician, arrived at the scene shortly thereafter to open the door considerably wider.

Vinogradov not only proved brilliant theorems, he also invented novel and brilliant methods to do so – the hallmark of a great mathematician. His methods were so powerful that, even though they were developed almost a century ago, some of them are still state of the art today. Vinogradov took another large step in the Goldbach Conjecture quest by showing that every sufficiently large odd integer was the sum of three primes.

What does a mathematician mean by the phrase "sufficiently large"? In the context of Vinogradov's result, it means that there is some integer N such that every odd number greater than N can be written as the sum of three primes. Now, since Vinogradov doesn't specify what the value of N is, it may not seem that this is a giant leap forward. Possibly, N is something like 1,000,000 – in which case, there's a lot of work to be done, but we can do it, or possibly 10 is a googol. (For those who think that googol is something to do with the search engine, the word "googol" was invented many years ago to describe the number 10^{100} – a 1 with a hundred zeros after it.) If that were the case, we'd probably never be able to do all that work, but it least it shows that the problem is no longer in the realm of the infinite but has been reduced to checking a finite number of cases.

Vinogradov's work rekindled interest in the Goldbach Conjecture, and as often happens in mathematics, new developments happened relatively rapidly – at least when compared to the hundred and fifty years or so after the formulation of the Goldbach Conjecture, when essentially nothing happened. And, as often happens in mathematics, when developments occur, they don't always occur in an expected direction. Seemingly, there were two different directions to advance existing results. One might either improve Schnirelmann's result by lowering the 800,000 number or hope to improve Vinogradov's result. This could possibly be done by finding an exact value for his "sufficiently large" or maybe by showing that the result he had obtained for odd integers could somehow be extended to even ones as well.

In 1948, Alfréd Rényi improved on Vinogradov's results. He showed there was an integer K such that every sufficiently large (that phrase again!) even integer was the sum of just two numbers! One of those numbers was a prime, and the other was almost a prime – it was a product of at most K prime factors. Twenty-five years later, Chen Jingrun was able to prove that every sufficiently large even integer was the sum of two numbers. Those two numbers were either both primes (that would be the case in which Goldbach was interested) or a prime and the product of two primes – Rényi's result with $K = 2$. To date, this is the best result in that direction.

Later Developments

Remember that concept of density we talked about earlier? In 1975, just half a century ago, Hugh Montgomery and Robert Vaughan showed that the set of even numbers which were not the sum of two primes was a set of density zero. If they had been able to show that the set of even numbers which were not the sum of two primes had a density greater than zero, it would have shown that the Goldbach Conjecture was false. Because the null set – the set with nothing in it – has a density of zero, their result meant that the Goldbach Conjecture could still be true.

And other mathematicians were working on whittling down Schnirelmann's 800,000 primes – and doing quite a good job of it. In 1995, Olivier Ramaré was able to show that every even number was the sum of not more than 6 primes. It's a long way from 800,000 to 6, and in 2013, Harald Helfgott showed that every even number is the sum of 4 primes.

That's where we stand now. It shouldn't be so difficult to go from 4 to 2, should it? Shouldn't we know shortly that the Goldbach Conjecture is either true or false, and we can then move on to other questions?

Sadly, there are two flies in this particular ointment. The first is that, even though we might be able to prove or disprove the Goldbach Conjecture, taking that last step might be really difficult. There's just no way to know. The second fly in the ointment was placed there by Kurt Gödel when he discovered that certain statements in mathematics are undecidable, i.e. not capable of being proved.

Could the Goldbach Conjecture be undecidable? Not only is there no way to know, there may never be a way to know. Just like there is no bell that rings when the stock market hits a bottom, there is no way to tell in general that a particular proposition is undecidable. There are a few propositions that are known to be undecidable, but they are artificial in the sense that they are propositions that have been constructed for the purpose of constructing an undecidable proposition. Some future mathematician may be able to show that the Goldbach Conjecture is undecidable, but who wants to do that? If you're going to spend time and energy working on the Goldbach Conjecture, there's a lot bigger pot of gold (metaphorically speaking) at the end of the rainbow if you can show that the Goldbach Conjecture is true or false. You'll get your name into the record books. If you show the Goldbach Conjecture is undecidable, you'll probably get a mention in works on mathematical logic. You'll probably also get a lot of the mathematicians who have been working on the Goldbach Conjecture hating you for telling them that they've been drawing dead all these years.

Could the Goldbach Conjecture Affect Your Life?

In some sense, it already has. It gave me an opportunity to talk about *The Devil and Simon Flagg*, and as you know, I consider that to be the most entertaining story about mathematics ever written. But I'm talking about how it could affect your day-to-day life.

Remember that Hardy felt that all the work he had done during his lifetime had been about the quest for truth and beauty in numbers. He would probably have been stunned to find out that the RSA algorithm, coupled with the difficulty he had proven regarding certain factorization problems, formed an integral part of the life of almost everyone who relies on passwords to protect information – and that's basically everyone.

The RSA algorithm relies on the difficulty of factoring a number that is the product of two primes. Showing that the Goldbach Conjecture is true might enable an algorithm that would be much more difficult to "break" than the RSA algorithm, which has already been shown to be vulnerable to quantum computing. Let's take a look at why this might be the case.

How long a list of primes are needed in order to be sure of finding two primes whose product is a given number? That's how sure we could be of breaking the RSA encryption. We need to examine all the primes less than the square root of n, and according to the prime number theorem, that's approximately $\frac{\sqrt{n}}{\ln(\sqrt{n})} = \frac{2\sqrt{n}}{\ln(n)}$.

Let's compare this with the length of the list of primes if the Goldbach Conjecture were true. We need to examine all the primes less than half of a given number. According to the prime number theorem, that's approximately $\frac{n/2}{\ln(n/2)} = \frac{n}{2(\ln n - \ln 2)} \approx \frac{n}{2\ln(n)}$ for large n. If you look at the ratio of the Goldbach number to the RSA number, it's on the order of the square root of n. This means that, for every prime the RSA algorithm looks at, an algorithm based on the Goldbach Conjecture might need to look at the square root of n primes.

But isn't it hard to manufacture such large primes? It certainly used to be, as the largest primes that were manufactured were known as Mersenne primes. A Mersenne prime has the form $2^p - 1$, where p is itself a prime. However, not every prime number p, when plugged into this formula, generates a prime number. For $p = 2, 3, 5$, and 7, $2^p - 1$ is prime, but $2^{11} - 1 = 2047 = 89 \times 23$.

However, times change, and although there are still people interested in finding ever larger Mersenne primes, from the standpoint of an encryption algorithm, it's no longer necessary. Remember that polynomial of degree 25 in 26 variables we mentioned earlier? You can plug in any collection of integers for each of the 26 variables and generate a prime as large as you want. Get two of them, add them up, and you have an even number.

So who knows? Maybe, at some point, you will rely on the Goldbach Conjecture for the security of your passwords, just as you rely today on the RSA algorithm. Or maybe quantum computers will become so good that they can break even a Goldbach Conjecture algorithm, and we may have to come up with a quantum algorithm in order to achieve password security that quantum computers cannot break. Or there may be something entirely different. Mathematics has a way of surprising us, and only time will tell.

Chapter 4

Fermat's Last Theorem

Nothing boosts the profile of a product or an event like having a celebrity spokesperson, so why should a mathematical quest be any different? And fortunately for us, we have a celebrity spokesperson who is known and beloved by hundreds of millions all over the world to endorse this particular quest. And fortunately for me, I don't have to pay him a cent because he's already done it.

The opening scene of the episode *The Royale* in *Star Trek: The Next Generation* finds Jean-Luc Picard in his quarters when Riker, his Number One, enters to tell him of an interesting development. Riker asks what Picard is doing, and Picard mentions that he's working on Fermat's Last Theorem. Riker, who wasn't big on math when he was at Starfleet Academy, inquires why, to which Picard replies that he finds it fascinating and adds that "in our arrogance, we feel we are so advanced and yet we cannot unravel a simple knot tied by a part-time French mathematician working alone without a computer".

So, who was this part-time French mathematician, and what is Fermat's Last Theorem?

Pierre de Fermat

Picard was right. Although Fermat did make contributions to both number theory and calculus, he was a lawyer and a government official. Fermat may have been the first person to whom the phrase "The reports of my death have been greatly exaggerated" applies, as he was a victim of the plague that swept through France in the early 1630s and was initially reported as being among the fatalities. We have the following as confirmation in a letter written by an acquaintance:

> *I informed you earlier of the death of Fermat. He is alive, and we no longer fear for his health, even though we had counted him among the dead a short time ago.*

Fortunately for Jean-Luc Picard – and for us – Fermat survived.

Long before Pythagoras, the Egyptians knew of the existence of the 3-4-5 right triangle. Even if you don't know the Pythagorean Theorem – and we can be certain that the Egyptians didn't – it's possible to see that a 3-4-5 triangle is a right triangle simply by placing them back-to-back, as shown in Fig. 4.1.

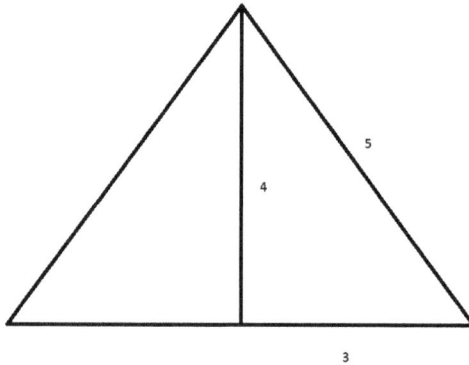

Fig. 4.1.

The numbers 3, 4, and 5 constitute a Pythagorean triple, but so do all integer multiples of 3, 4, and 5, such as 6, 8, and 10 or 30, 40, and 50. But these triangles are similar to the original 3-4-5 right triangle; they have exactly the same shape. To this day, carpenters who haven't got a T-square handy for constructing right angles are advised to make a 3-4-5 triangle.

The Greeks knew there were an infinite number of Pythagorean triples representing an infinite number of differently shaped right triangles. Euclid supplied a simple proof of this fact. Observing that $(n+1)^2 = n^2 + (2n+1)$, all that would be needed was for $2n + 1$ to be a square, say $2n + 1 = k^2$.

We would then have $(n + 1)^2 = n^2 + k^2$. But $2n + 1$ runs through all the odd numbers as n goes from 1 to infinity, and since the square of any odd number is an odd number, we will see all the odd squares among the different values for $2n + 1$. For instance, if $2n + 1 = 49 = 7^2$, then $n = 24$, and we get the 7-24-25 right triangle.

You might ask if this proof gets us all the differently shaped Pythagorean triples, and the answer is that it doesn't, as is evidenced by the 8-15-17 right triangle. And so, the next question a mathematician would ask is, "Can we find a formula which generates all the differently shaped Pythagorean triples?"

From the standpoint of discussing Fermat's Last Theorem, this is certainly a digression. But that's part of the appeal of a mathematical quest: You can start out along one path and be diverted by something bright and shiny, which takes you in a totally different direction.

But it's time to return to what is unquestionably the best-known quest in mathematical history. The reason that we spent a little time with Pythagorean triples is that many questions in mathematics result from attempts to extend known results, that is, apply them to a larger collection of situations than the original. Fermat was obviously wandering down this road when he scribbled the following in a math book that he was reading:

> To divide a cube into two cubes, a fourth power, or in general any power whatever into two powers of the same denomination above the second is impossible, and I have assuredly found an admirable proof of this, but the margin is too narrow to contain it.

As far as we can tell, Fermat never wrote down the proof to which he referred. He did, however, prove this theorem for $n = 4$ (the theorem that there are no integer solutions to $a^4 + b^4 = c^4$). As one who has been there and done that (by this I mean believing that a proof that handles one case handles all cases), I'm guessing that he thought the methods he used to prove the special case of $n = 4$ would suffice for any arbitrary n.

And that completed the seventeenth century's contribution to Fermat's Last Theorem. But in the eighteenth century, Leonard Euler, unquestionably one of the greatest mathematicians of that or any other century, managed to prove the special case of $n = 3$. Two cases down – an infinite number still to go.

This is how mathematics gets done. When you read a math textbook, it's presented for you in a neat and orderly fashion – a logical development that seems very straightforward when you read it. But mathematics is no different from any other creative endeavor. We have Beethoven's sketchbooks to show how diligently he worked to produce his symphonies; the story is pretty much the same for Shakespeare's plays or Monet's *Water Lilies*.

Chipping Away

Progress on Fermat's Last Theorem started to come more quickly in the nineteenth century. Two eminent mathematicians, Adrien-Marie Legendre and Johann Dirichlet, managed to nail the $n = 5$ case, A lesser-known mathematician (at least to me, I knew of both Legendre and Dirichlet but not this next guy), Gabriel Lamé, managed to prove Fermat's Last Theorem for $n = 7$.

The first mathematician since Fermat to utter "eureka" for the Last Theorem was Ernst Kummer, who announced a proof of the theorem in 1843. Unfortunately – at least for Kummer – the proof did not stand up, as flaws were found in the proof. Kummer did, however, make a notable dent in the Last Theorem. Up until Kummer, the only results that had been shown to be valid were one-offs, for $n = 3, 4, 5$, and 7. But Kummer managed to show that the Last Theorem was true for all regular primes.

You might hope that the definition of "regular prime" would be something that could be easily understood. Sadly, that's not the case. Shamefaced confession: I looked at the definition of regular prime, and I didn't understand it.

Maybe Kummer himself didn't completely understand it! His definition of regular prime was so difficult to grasp that Kummer wrote a letter to a fellow mathematician saying that he didn't think 37 was a regular prime.

But now we know – at least as far as 37 is concerned. I did find a list of the first few regular primes, so I present for your enjoyment the regular primes less than 100:

$3, 5, 7, 11, 13, 17, 19, 23, 29, 31, 41, 43, 47, 53, 61, 71, 73, 79, 83, 89, 97.$

In fact, if you look at this list, the only primes less than 100 that are not regular are 37, 59, and 67. So, whatever they are, it seems like there are probably a lot of them, right? Most of the primes less than 100 are regular – whatever that is.

After all, Euclid proved that there were an infinite number of primes with a simple and ingenious proof, which we saw in the previous chapter. As of this writing, no one has been able to determine whether or not there are an infinite number of regular primes. Nonetheless, Kummer's result was a great step forward for two reasons. First, he was able to show that Fermat's Last Theorem was true for a bunch of numbers all at once, rather than simply proving it for one value of n at a time, as Fermat, Euler, Legendre, Dirichlet, and Lamé had done. Second, he had come up with arguments that had never been used before.

New proof techniques are the lifeblood of mathematics. We've already seen one example of this: the transformations given by the Italian algebraists. We'll see – or at least reference – several more of these throughout the book.

There are a *lot* of unproven results in mathematics. But proving a new result simply adds one more brick to the mathematical edifice; coming up with a new proof technique, however, gives mathematicians a new tool and possibly an important one.

Kummer had come up with a new tool, but it wasn't good enough, in the words of Jean-Luc Picard, to "unravel a simple knot tied by a part-time French mathematician working alone without a computer" – not that they didn't try. In the four years between 1908 and 1912, over 1,000 erroneous proofs of Fermat's Last Theorem were actually published.

I must admit, when I saw that number, I was shocked and, also, a little incredulous, so I decided to do a little arithmetic. A typical mathematics journal publishes monthly and has maybe 10 articles. I don't know how many journals were around in 1908, but I think 100 would be a generous estimate. That means 1,000 articles were published every month, so in the four-year period between 1908 and 1912, maybe 1 of every 50 articles was a false proof of Fermat's Last Theorem.

You might be curious as to how false proofs get published – didn't somebody check these things? Yes, they do. Most articles are sent to a referee – someone with knowledge of the field – who reads them and does his or her best job of trying to decide whether the results are correct and whether the contribution is worthwhile. But referees are humans, so they make mistakes. But a mistake in a proof of Fermat's Last Theorem doesn't have anywhere near the same catastrophic potential as if the Food and Drug Administration certifies a drug to be safe that turns out to have serious side effects. So, there will be a number of journal articles that will be published that have mistakes, but sooner or later someone will either notice that there is a mistake or use the result to prove something that is demonstrably false. When that happens, there must have been a foul-up somewhere earlier in the chain that led up to the demonstrably false result, and the error will be found.

And so, with one thing and another, virtually no progress was made on Fermat's Last Theorem until the middle of the twentieth century, when a Japanese mathematician discovered a connection between a class of curves and Fermat's Last Theorem.

Ellipses and Elliptic Curves

If you've taken a course in analytic geometry, you've probably studied conic sections. These curves are so-called because they represent the intersection of a plane with a cone. One of those curves is an ellipse, which can also be characterized as the set of all points for which the sum of the distances from two given points is a constant.

This definition is much more useful than the one about the intersection of a plane with a cone because we can bring the powerful tool of analytic geometry into the picture. And analytic geometry is an immensely powerful tool – far more than (dare I say it?) the geometry of the Greeks. It melds geometry with algebra to the benefit of both disciplines.

Analytic geometry enables us to visualize and describe properties of curves that were simply unknown to the Greeks. During the middle of the twentieth century, it was discovered that Fermat's Last Theorem, which belongs to the mathematical discipline known (not surprisingly) as number theory, had deep connections to a class of curves which are represented using analytic geometry by the equation $y^2 = x^3 + Ax^2 + Bx$, where A and B are integers. These curves are known as elliptic curves. Figure 4.2 depicts an ellipse side-by-side with an elliptic curve.

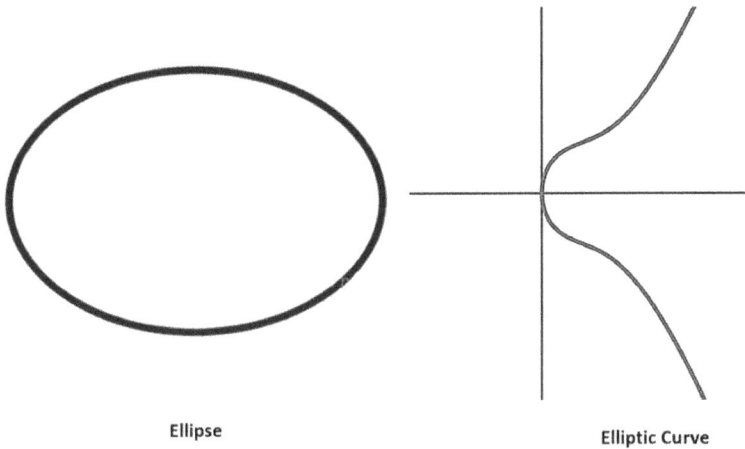

Ellipse

Elliptic Curve

Fig. 4.2.

See the similarity? Frankly, I don't either, but I'm sure there's a reason why elliptic curves are called elliptic curves. At least, it's almost certainly a better reason than why Kummer's class of primes are called regular primes. If Kummer himself couldn't tell you whether 37 was a regular prime, how on earth could they be regular?

Yutaka Taniyama

This paragraph condenses something that I not only don't understand, but I'm also willing to bet it would take me several months of study to get to the point where I understand what people are talking about. Nonetheless, I get the general idea, and I'm pretty sure you will, too. Some areas of mathematics have been intensively investigated, and there are a lot of theorems and proof techniques available. Others, not so much. While progress on Fermat's Last Theorem had virtually ground to a halt (despite the thousand or so erroneous proofs that were published between 1908 and 1912), significant progress had been made in the study of elliptic curves. And in 1955, the Japanese mathematician Yutaka Taniyama managed to show that if one could establish a particular result concerning elliptic curves, one could prove Fermat's Last Theorem. This connection was embellished by Goro Shimura and became known as the Taniyama–Shimura Conjecture.

One would think that when your work has attracted sufficient atten- tion that something gets named after you and people are avidly pursuing the path you have outlined, your future is bright. However, in a stunning development, three years after the Taniyama–Shimura Conjecture was for- mulated, Yutaka Taniyama committed suicide. There was no forewarning of this shocking event. He had met his fiancée Misako in November 1957, and they were quite far along in their plans. They had signed a new lease and even purchased kitchen utensils. His suicide note ends as follows:

> *Merely may I say, I am in the frame of mind that I lost confidence in my future. There may be some to whom my suicide will be troubling or a blow to a certain degree. I sincerely hope that this incident will cast no dark shadow over the future of that person. At any rate I cannot deny that this is a kind of betrayal, but please excuse it as my last act in my own way, as I have been doing all my life.*[1]

One person over whom this event cast a dark shadow was his fiancée, Misako. She committed suicide a month later, writing:

> *We promised each other that no matter where we went, we would never be separated. Now that he is gone, I must go too in order to join him.*

I said earlier in this section that it would take me several months of study to reach the point where I understood enough about elliptic curves to understand the Taniyama–Shimura Conjecture. But it would take me a life- time of study to understand why Yutaka Taniyama committed suicide with such a shining future, both personal and professional, ahead of him. And it would take another lifetime to see how an individual with the intellectual brilliance to capture the connection between elliptic curves and Fermat's Last Theorem could not see the effect that his suicide would have on his fiancée.

---

[1] https://mathshistory.st-andrews.ac.uk/Biographies/Taniyama/.

The Frey Curve

Two important developments occurred during the 1980s. The first was the association made by Gerhard Frey between a particular elliptic curve and Fermat's Last Theorem. That curve is now known as a Frey curve and has the form $y^2 = x(x - a^n)(x + b^n)$, where a, b, and n are integers.

You don't have to be a specialist in elliptic curves to see that the quantities a^n and b^n in the Frey curve are the same fellows that show up in the $a^n + b^n = c^n$ formulation that is central to Fermat's Last Theorem. You do, however, have to be a specialist to understand what Frey enunciated – that if numbers a, b, c, and n existed satisfying the Fermat relationship, there would be an elliptic curve that lacked a certain property (for the curious, that property is known as modularity). So, to the complete surprise of anyone who had been working on Fermat's Last Theorem for the first three hundred years of the conjecture's existence, there was a road to a proof that led through the properties of certain geometric curves.

And the second important development that occurred during this generation was that Andrew Wiles, a talented British mathematician, had decided to attack the problem – and he did so in complete secrecy.

Andrew Wiles

Wiles had become fascinated with Fermat's Last Theorem when he read about it as a 10-year-old. He had found it amazing that he could easily understand what the theorem said but that no one had been able to prove it. He wanted to be the first to prove it. However, by the mid-1980s, he was already an established mathematician with stellar credentials, working in areas that were not that far removed from Fermat's Last Theorem. When he took note of the progress that had been made during the 1980s, he decided to try to fulfill his childhood dream.

Telling only his wife about this particular quest, he decided to devote an hour a day to the project. And, six years after he began his particular quest, he gave a series of three lectures entitled Modular Forms, Elliptic Curves and Galois Representations. Nobody who initially attended had any idea of the bombshell that Wiles intended to detonate. But, at the conclusion of the third lecture, he stated that he had proved a general case of the Taniyama–Shimura Conjecture. He concluded his lecture by saying that this also proved Fermat's Last Theorem.

An announcement of the proof of a general case of the Taniyama–Shimura Conjecture would undoubtedly never have come to public attention. But the announcement of a proof of Fermat's Last Theorem, on the other hand, was front-page news.

Extraordinary claims require extraordinary proof – and that certainly goes for a claim that you have just proved Fermat's Last Theorem. Wiles' proof was subject to close scrutiny by everyone in the mathematical community with expertise in elliptic curves, and holes began to appear in the proof.

Patching Up the Proof

When holes begin to appear in your roof, you'd better get it patched up before the next rainy season. And when holes begin to appear in your proof, a patch job is also in order.

The mathematical community is now considerably different from what it was in the eighteenth century, when Paolo Ruffini was not notified of a flaw in his proof of the inability to solve the quintic by radicals. It is vastly different than it was in the sixteenth century, when mathematicians would conceal what they knew to use as future weapons in intellectual duels. Today, once the mathematical community senses an important advance is on the horizon, it generally works together to ensure that advance.

Such was the case in the early 1990s, when holes began to appear in Wiles' initial proof. Time and again it would seem that the holes had been patched, only for another hole to appear – sort of like Whack-a-Mole for Fermat's Lat Theorem. The community sensed that a proof was close, but it was not there yet. As André Weil, one of the *éminences grises* in the mathematical community stated:

> *I believe he has had some good ideas in trying to construct the proof but the proof is not there. To some extent, proving Fermat's Theorem is like climbing Everest. If a man wants to climb Everest and falls short of it by 100 yards, he has not climbed Everest.*

More than a year went by. But then, one day, Wiles had an inspiration. As he stated:

> *It was so indescribably beautiful; it was so simple and so elegant. I couldn't understand how I'd missed it and I just stared at it in disbelief for twenty minutes. Then during the day I walked around the department, and I'd keep coming back to my desk looking to see if it was still there. It was still there. I couldn't contain myself, I was so excited. It was the most important moment of my working life. Nothing I ever do again will mean as much.*[2]

You don't need to understand any of the details of what Wiles was studying – all you need to do is see the sentence "It was so indescribably beautiful; it was so simple and elegant" to appreciate what it is like when a centuries-long quest ends at the intersection of truth and beauty.

Flagg Redux

Wiles' initial announcement and the final completion of the proof took place in 1993–1994. Probably, 99% of the mathematicians in the world were like me in that they had never worked on, or anywhere near, Fermat's Last Theorem. But absolutely 100% knew of Fermat's Last Theorem and could appreciate what a great result this was.

And this was during the period when I approached a publisher with my idea of a liberal-arts math textbook which would lead off with Arthur Porges' wonderful story *The Devil and Simon Flagg*. Porges, however, had initially published the story in the early 1950s, and the problem that Simon Flagg posed to the Devil was Fermat's Last Theorem.

---

[2]https://mathshistory.st-andrews.ac.uk/Biographies/Wiles/.

But when I started writing the book, Wiles – with the help of the mathematical community – had completed the proof of Fermat's Last Theorem. Porges was aware of this when I initially contacted him, and he had no problem assenting to the idea that the story should be changed, with Flagg posing the question of the Goldbach Conjecture to the Devil. As I mentioned, Porges had spent some time as a mathematics instructor in California colleges and community colleges, and of course he was familiar with the Goldbach Conjecture. We changed a few minor details of the story to make it more up-to-date, as there were a bunch of references to everyday life in the 1950s that needed to be replaced, but the major change was to replace the question that Simon posed to the Devil. We have seen that Wiles' proof went through several upheavals from the time of his initial announcement to the moment that the mathematical community decided that the work was complete. However, when the initial announcement was made, I was unaware that there were holes in the proof that needed patching. As a result, I suggested to Porges that we replace Fermat's Last Theorem with the Goldbach Conjecture. The story would lose a lot of its punch if the Devil were given a problem that was already solved.

Wiles first came across Fermat's Last Theorem when, as a ten-year-old boy, he read about it in Eric Temple Bell's *Men of Mathematics*. But I'd like to think that his interest in the problem got rekindled sometime in the 1980s by reading *The Devil and Simon Flagg*.

Part III

Recent – and Relatively Recent – Quests

I'm sure historians have a fairly accurate idea of when the Renaissance ended and what we might call the modern era began. Possibly, historians of mathematics do as well, but I admit I tend to think of modern mathematics as beginning when Newton and Leibniz developed calculus. Calculus was a game-changer – quite literally, as it enabled mathematicians to deal with changing quantities. The Greeks could handle many computations, such as areas and volumes, but the objects in question remained fixed and unchanging. The Greeks could certainly have handled simple problems such as how far a man would walk in four hours if he walked at a rate of three miles per hour (or whatever units the Greeks used to measure distance and time). But the glory of calculus was that it could answer such questions if the man was walking more slowly at some times and more rapidly at others, as well as a host of questions that developing technology and an expanding science required.

But the quests started coming thick and fast with the nineteenth century and the expansion of the field of mathematics. If one looks at a standard college mathematical curriculum, three semesters of calculus are generally taught in the first two years, and most of this was completed in the seventeenth and eighteenth centuries. But the upper division courses in algebra, probability, analysis, and topology didn't really begin until the nineteenth century, which also saw the beginning of mathematical logic beyond what Aristotle, Venn, and Boole had developed. So, the modern quests start here.

Some of these quests have been resolved, but some are still ongoing. And it's virtually certain that even more were just recently begun or have yet to begin. There are more mathematicians alive now than ever before, and the ability to communicate with other mathematicians is far greater than it has ever been.

And that's how quests frequently start.

Chapter 5

The Incompleteness Theorem and the Continuum Hypothesis

While the bone and muscle of mathematics may be its ability to calculate, the heart and soul of mathematics is its ability to assert beyond all possible doubt that a particular conclusion is true.

Take the following well-known result from geometry: The sum of the angles in a triangle is $180°$. It's not always true; the following is a picture of a triangle with three right angles, totaling $270°$ (Fig. 5.1).

As you can tell, though, this triangle is located on a sphere rather than a plane. Geometry on a sphere is different from geometry on a plane. This certainly doesn't cause us to question the validity of the theorem that the sum of the angles in a triangle is $180°$; it makes us realize that the axioms for geometry on a sphere must be different from those for geometry on a plane.

For millennia, mathematicians had operated under two tacitly understood assumptions. First, if they made a mathematical statement, it was either true or false. They may not have been able to establish which of these two alternatives held, but it had to be one or the other. Second, they could rely on the methods of logic that had been codified by Aristotle.

Aristotle was probably the first individual to attempt a serious study of logic. To give you some idea of the brilliance of Aristotle's work, the great philosopher Immanuel Kant said that in the two millennia between Aristotle's time and his own, nothing significant had been added by anyone. And Aristotle's treatise on logic was encyclopedic; he appeared to cover all bases that needed to be covered. Kant certainly thought so.

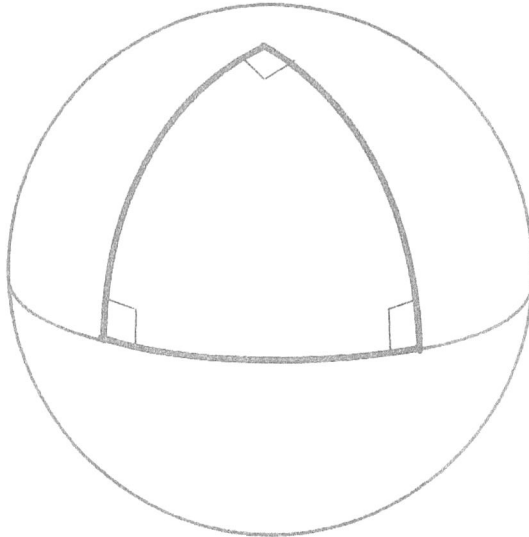

Fig. 5.1.

Still, there were unusual situations that enabled philosophers and theologians to weigh in. During the Middle Ages, there was much debate about the barber paradox: If the barber shaves everyone in the village, who does not shave himself and only those men, then who shaves the barber? Does the barber shave himself? If so, he's shaving someone other than the men who do not shave themselves. If not, then he isn't shaving everyone who does not shave himself.

The barber paradox has an antecedent that goes back to the Greeks and is known as the Epimenides Paradox. Surprisingly, we know almost nothing about Epimenides except this particular paradox; it's almost like the Cheshire cat, which is only known by its smile.

Let's put the paradox of Epimenides in the form of a four-word sentence: This statement is false. Here is a very simple statement that can neither be true nor false. If "this statement is false" is true, we must accept as true that it is false; in other words, it's both true and false. If "this statement is false" is false, then if we believe that "this statement is false" is either true or false and we have said that it is false, then it must be true. But then, once again, if "this statement is false" is true, then it must also be false.

It may seem that this particular paradox is simply a play with words, something with no deep significance. And although philosophers and theologians did spend some time wrestling with this, it played no role in the development of mathematics. It has nothing to do with arithmetic, algebra, geometry, calculus, or probability – the branches of mathematics that were being studied until perhaps the middle of the nineteenth century.

One of the reasons for this was that mathematics was concerned with what problems it could solve. Calculus, for instance, solved an amazing number of problems, and in order to use the tools of calculus, it wasn't really necessary to look deeply into the axiomatic structure on which the truths of calculus relied. The only really well-defined axiomatic structure was the one defined by Euclid for doing geometry. Most of the axioms and postulates made intuitive sense, but the parallel postulate – that given a line and a point not on the line, there was one and only one line through the point which would never meet the given line – had occasioned some thought. And by the middle of the nineteenth century, geometric structures had been developed which did not satisfy this postulate. Mathematicians are fond of sticking the prefix "non-" in front of adjectives when they are confronted with the existence of objects that don't possess a desired property. Commutativity is a desirable property for an algebraic operation, but if you don't have it, no worries; you're just studying non-commutative algebraic structures, such as non-Abelian groups. The parallel postulate is a desirable property for certain geometric structures, such as the plane, but if you don't have it, no worries; you're just studying non-Euclidean geometries. And mathematicians, like other scientists, love having new things to study.

Until the nineteenth century, the infinite was something toward which mathematicians adopted a somewhat standoffish attitude. Infinity could be approached but never realized. Since there were no infinities in the real world, there didn't seem to be a need to treat infinities as a mathematical quantity.

During the middle of the nineteenth century, mathematicians had managed to come up with reliable demonstrations of how the rational numbers were constructed from the integers and how the real numbers were constructed from the rational numbers. But until the 1880s, no one had managed to construct a good axiomatization for the integers. As Richard

Dedekind so famously said, the integers were the work of God, and all else was the work of man. But it is notoriously difficult to understand the work of God.

Nonetheless, in the 1880s, the Italian mathematician Giuseppe Peano had managed to come up with an axiomatization for the integers. It was at about this time that mathematicians started to examine the question of what made for a "good" mathematical system and what didn't. The idea they came up with was consistency. A system is consistent if no statement that can be made within the system is both true and false. One certainly doesn't want to work with a system in which the same proposition is both true and false, "true" and "false" being inherently incompatible concepts.

Proving that a system is consistent seemed to be a lot harder than proving theorems in an axiomatic system. The first two decades of the twentieth century saw the publication of Alfred North Whitehead and Bertrand Russell's titanic work *Principia Mathematica*, which was a serious examination of the foundations of mathematics. It took more than 300 pages for Whitehead and Russell to prove the statement that $1 + 1 = 2$.

Principia Mathematica triggered a firestorm of activity. One of the mathematicians inspired by this work was Emil Post, who was able to show in his doctoral dissertation that the formalization of the propositional calculus (the logical methods by which the truth or falsehood of propositions is deduced) in *Principia Mathematica* was consistent. At last, mathematics had something on which they could be certain they could rely.

But the consistency of integer arithmetic – the very foundation on which all of mathematics appeared to rely – had not yet been established. Normally, when one constructs a building, the first thing one does is put in a solid foundation. The first floor is built upon this solid foundation, the second floor upon the first, and so on. Mathematics had gone about things differently: It had built a solid foundation for the real numbers based on the rational numbers, and it had built a solid foundation for the rational numbers based on the integers. The solid foundation for the integers, however, was yet to be established.

One of the titans of nineteenth- and twentieth-century mathematics was David Hilbert. Hilbert was not just a mathematician; he was also an awesomely talented physicist – more about that in a subsequent chapter.

At the turn of the twentieth century, Hilbert posed twenty-three unsolved problems that he felt would significantly further the development of mathematics. Hilbert's Second Problem was to establish a consistent formulation for the integers so that the magnificent edifice which had been constructed would have a solid foundation on which to rely.

Meanwhile, mathematicians were beginning the process of exploring two new areas of study. The first was set theory, which had begun the task that mathematicians of the past had avoided – the study of infinite quantities. The second was the coming of the computer. True, by the first part of the twentieth century, the electronic computer had not yet made an appearance, but everyone knew it was only a matter of time. The computer promised to automate all those calculations that would have taken too long if done by hand. Many mathematicians, Hilbert included, wondered if it would be possible to automate the process of proving theorems.

The mathematicians who were preparing for the coming of the computer had realized that there was a potential problem. A successful computation would terminate with the computer spitting out the answer to a problem and then stop. However, it was possible that a badly written computer program would go into a repetitive loop in which the computer would perform the same computations over and over again without ever stopping. One way to prevent this from happening would be if it were possible to write a computer program which examined other computer programs with the goal of determining whether the examined program would halt or loop. This was the so-called halting problem.

The brilliant British mathematician Alan Turing, whose efforts were to prove pivotal in deciphering the German Enigma encryption device during World War II, came up with a brilliant solution to the problem. Inspired by the Epimenides Paradox, Turing showed that it would not be possible to write a computer program that could successfully examine other computer programs to determine whether they would halt or loop.

Here's how Turing's proof worked. Suppose that there really is a program, which we'll call HaltChecker, that can examine a program P and input I and determine whether P halts or loops using that input. We construct a new program, call it WrongAnswer, which is built upon HaltChecker but does exactly the opposite. WrongAnswer uses the HaltChecker inside it to examine a program P and input I, but if HaltChecker determines that P will halt on input I, WrongAnswer will loop; and if HaltChecker determines that P will loop on input I, WrongAnswer will halt.

Now we ask WrongAnswer to use itself as both program and input. If the inside HaltChecker determines that using WrongAnswer as both program and input will halt, then WrongAnswer will loop (that's what WrongAnswer does when the HaltChecker inside it tells it that the input program will halt). Similarly, if the inside HaltChecker determines that using WrongAnswer as both program and input will loop, then WrongAnswer will halt (that's what WrongAnswer does when the HaltChecker inside it tells it that the input program will loop). So in both cases, WrongAnswer is doing the reverse of what HaltChecker says it will do. This contradiction shows that it is impossible to write a HaltChecker program that can tell whether a program–input combination will halt or loop.

What Is a Math Conference Like?

I'm a huge tennis fan – and the four most important tennis tournaments of the year are called Grand Slams. There's one in England, one in France, one in Australia, and one in the United States. The two prestigious events in each are the Gentlemen's Singles and the Ladies' Singles. Each of the tournaments takes two weeks, and there are lots of courts available. By the end of the first week, so many competitors would have lost that smaller courts are used for conducting tournaments of lesser importance, such as the Mixed Doubles or the Under 14-Year-Olds.

And that's a lot like a big math conference. There are the prestigious events: the keynote addresses and the delivery of the important papers (generally those are identified by having the top names in the field, just like tennis); and then there are the lesser events, held far from the limelight.

That's what it was like at the Königsberg Conference in 1930. Although Hilbert wasn't there, his presence was felt, and many of the top mathematicians, such as John von Neumann, were in attendance. And far from the limelight, the as-yet-unknown mathematician Kurt Gödel would deliver a paper that would shake the foundations of mathematics.

Hilbert, like most of the other leading mathematicians of the day, believed that every theorem in mathematics was either true or false. He talked about the axiom of the solvability of every problem and declared that the conviction of the solvability of every mathematical problem is a powerful incentive to the mathematician. Hilbert was aware that there were important unsolved problems, but he felt that, given enough time and enough brilliance, they could all be solved.

Before Galois conclusively demonstrated the insolvability of the quintic, Ruffini and Abel had glimpsed the truth that the quintic could not be solved by radicals. But it doesn't seem that before Gödel, anyone dared to venture the thought that some problems simply might not be decidable. What Gödel was able to show, using a variation of the Epimenides Paradox, that if the Peano Postulates were consistent (and no one thinks otherwise, even though it has yet to be shown), there were theorems in arithmetic that could not be proven either true or false using the tools of arithmetic.

That doesn't mean that these theorems were neither true nor false. The Goldbach Conjecture is a good example. If we had an infinite amount of time, we could examine every even number and see whether or not it is the sum of two primes: $4 = 2 + 2, 6 = 3 + 3, 8 = 3 + 5, 10 = 3 + 7$, etc. But we don't have an infinite amount of time. In order to settle the Goldbach Conjecture, we must do one of three things: We must prove that every even number is the sum of two primes, show that the assumption that every even number is the sum of two primes leads to a contradiction, or find an even number which is not the sum of two primes. Gödel did not exhibit a well-known conjecture for which this could not be done; the theorem that he constructed was highly artificial and would not be the type of theorem on which any mathematician would waste his or her time. But it was a theorem of arithmetic, and Gödel was able to show that given the rules for arithmetic proof, it was impossible to establish that the theorem was either true or false.

Gödel is yet another example, like Taniyama, of how mathematical brilliance is no armor against psychological onslaughts. Gödel moved from Europe to the United States in order to escape the Nazi threat and eventually ended up at the Institute for Advanced Studies in Princeton, New Jersey. Never an extrovert, he began to retreat from contact with other human beings. At one time, the only person with whom he would converse, other than his wife, was Albert Einstein, who also had taken residence in Princeton and was also working at the Institute for Advanced Studies.

In his later years, Gödel developed extreme paranoia and became convinced that other people were trying to poison him. He would only eat meals prepared by his wife, and after she passed away, Gödel starved to death.

The Continuum Hypothesis

Not only was arithmetic proving a whole lot trickier than mathematicians had heretofore believed, set theory was as well.

Just as the Peano Postulates had become the gold standard for axiomatic systems for arithmetic, a system of axioms known as "ZFC", which stands for Zermelo–Fraenkel plus the Axiom of Choice, had attained a similar position in set theory. If you examine the Zermelo–Fraenkel axioms, most of them seem fairly obvious: Two sets are equal if and only if they contain the same things and others of a similar nature. However, closer inspection revealed that something was missing.

Many set-theory proofs have a statement on the order of "choose an element x from set A". Though seemingly pretty straightforward, how do you know you can do that? Bertrand Russell came up with a beautiful way to illustrate the dilemma. Suppose your set consists of a pair of shoes. You can say, "Choose the left shoe", because there is something intrinsic to shoes (left and right) that enables you to choose one. But what if your set consists of a pair of socks? There's nothing intrinsic to socks that enables you to choose one. So, in the absence of being able to specify a rule to enable one to choose an element from a set, it was necessary to attach an axiom – the Axiom of Choice – which allowed you to make this choice.

Set theory is a branch of mathematics that was mostly nonexistent until Georg Cantor showed how to deal with infinities. Cantor's brilliant insight was the idea of cardinality: The way you classified sets was by putting them in one-to-one correspondence with certain standardized sets. Any set with cardinality three, for example, can be put into one-to-one correspondence with the set consisting of the numbers 1, 2, and 3, or with a set of three cookies, or three stars.

Among Cantor's achievements were to show that there were different infinities – at least, when viewed from the standpoint of cardinality. In a truly brilliant proof, Cantor showed that the set of real numbers between 0 and 1 could not be put in one-to-one correspondence with the set of all integers. He did this by what is now known as a diagonal argument, and it's not at all hard to see how it works.

Assume that we actually can put the set of all real numbers into one-to-one correspondence with the integers. That means we could make a list of all real numbers, with r_1 being the first number on the list, r_2 being the second number on the list, r_3 being the third number on the list, r_4 being the fourth number on the list, etc.

Any real number between 0 and 1 has a decimal expansion. We'll write out what the decimal expansion of the first three numbers on the list look like:

$$r_1 = 0.d_{11}d_{12}d_{13}d_{14}\ldots,$$

$$r_2 = 0.d_{21}d_{22}d_{23}d_{24}\ldots,$$

$$r_3 = 0.d_{31}d_{32}d_{33}d_{34}\ldots,$$

$$r_4 = 0.d_{41}d_{42}d_{43}d_{44}\ldots,$$

and so on. d_{24}, for example, is the fourth digit after the decimal point of the second number, r_2, on the list. By the assumption that the real numbers between 0 and 1 and the integers can be put in one-to-one correspondence, the list contains *every* real number.

And here's where Cantor did something unbelievably brilliant. He constructed a number that was not on the list. My wife's birthday is September 1, so to honor her, I'll show how Cantor did it using the digits 1 and 9.

The following is a copy of the decimal expansion that you see above, except I'm going to bolden the first digit of the first number, the second digit of the second number, the third digit of the third number, and the fourth digit of the fourth number:

$$r_1 = 0.\mathbf{d_{11}}d_{12}d_{13}d_{14}\ldots,$$

$$r_2 = 0.d_{21}\mathbf{d_{22}}d_{23}d_{24}\ldots,$$

$$r_3 = 0.d_{31}d_{32}\mathbf{d_{33}}d_{34}\ldots,$$

$$r_4 = 0.d_{41}d_{42}d_{43}\mathbf{d_{44}}\ldots$$

You can see that the digits in bold form the diagonal of the square, going from the upper-left corner downward to the right – that's where the term "diagonal argument" comes from.

Here's how Cantor constructed a number r that was not on the list. He did so digit by digit. If $d_{11} = 1$, the first digit of r was 9. If $d_{11} \neq 1$, then the first digit of r was 1. So, r has a different first digit from r_1.

If $d_{22} = 1$, the second digit of r was 9. If $d_{22} \neq 1$, then the second digit of r was 1. So, r has a different second digit from r_2.

You can see how this works, but we'll do it for one more digit. If $d_{33} = 1$, the third digit of r was 9. If $d_{33} \neq 1$, then the third digit of r was 1. So, r has a different third digit from r_3.

And so on. Because r and r_n have different *n*th digits, r does not appear anywhere on the list.

Cantor, who spent a portion of his life in and out of the mental institutions of the day, was indeed the master of the infinities. Using a different diagonal argument, he was able to show that if one had a set of a particular cardinality, the set of all subsets of that set could not be put in one-to-one correspondence with the original set. For instance, the set of all sets of integers cannot be put in one-to-one correspondence with the integers.

But one thing Cantor – or any other mathematician – had not been able to do was determine whether there was a set with a cardinality between the cardinality of the integers and the cardinality of the set of all real numbers in the closed interval [0, 1]. There were plenty of interesting candidates. The rational numbers had been shown to have the same cardinality as the integers, of course, by one of Cantor's diagonal arguments. The same for the algebraic numbers – all those numbers which were roots of polynomials. The transcendentals – numbers such as pi which were not the roots of polynomials – had the same cardinality as that of the continuum.

After several failed attempts, Cantor suggested in 1878 that every subset of real numbers either had the same cardinality as the integers or the same cardinality as the continuum. This conjecture became known as the continuum hypothesis, and Hilbert felt that establishing the truth or falsehood of the continuum hypothesis was so important that he made it the first of his Twenty-Three Problems.

Nothing significant was done on the continuum hypothesis for decades. Then, in 1940, Kurt Gödel showed that the negation of the continuum hypothesis – that there was a set of real numbers with a different cardinality than that of the integers or that of the continuum – could not be proved in ZFC. Then, in 1963, Paul Cohen constructed a new type of argument he called "forcing" (which is beyond the scope of this book to discuss), in which he was able to show that the continuum hypothesis itself could also not be proved in ZFC. Logicians describe this by saying that the continuum hypothesis is independent of ZFC. It actually needs to be a separate axiom from ZFC, and it is possible to construct mathematical systems for two different extensions of ZFC, one in which the continuum hypothesis is true and one in which the continuum hypothesis is false.

How the Independence of the Continuum Hypothesis Changed My Life

In 1963, I was just starting graduate school and was blissfully unaware even of the existence of the continuum hypothesis. When I began work on my doctorate, my adviser had me work on a problem from the area in which he was currently working. This is almost certainly the way the vast majority of doctoral students earn their degrees: by working on problems for which their advisers have substantial expertise.

The particular area in which my adviser worked had a conjecture, known as "uniqueness of norm", which had not yet been resolved. Most of the researchers in the area felt that it was true, as there were a number of results which came close to establishing it, but no one had nailed it down. Much of the work I did between finishing graduate school and 1980 was on this problem or ones associated with it. I remember one time in the 1970s when I felt I had solved this problem, wrote up my solution, and was checking it one more time before submitting it to a journal – when I discovered that my proof used a sneaky variation of the obviously invalid assumption that $0 = 1$. I consoled myself with two thoughts, the first being that I hadn't mailed the article and the second being that I'd saved money because I hadn't even put stamps on the envelope yet. But I continued to be intrigued by it and to work on it.

But in 1980, my father needed advanced medical care that insurance did not cover and that the salary of an associate professor (which I was at the time) was insufficient to defray. Fortunately, one of my friends had become an options trader, and for five years, I made my living – a much more lucrative one than teaching mathematics – trading stock options. But when the necessity of making enough money to cover medical costs had passed, I wanted to get back to teaching and doing math. It's hard to beat a job in which people pay you to teach and think about something you love.

During the five years that I was a stock options trader, I paid no attention to what was happening to the uniqueness of norm problem. During those five years, an absolutely stunning development had taken place. It had been shown that the resolution of the uniqueness of norm took two different paths.

Earlier, I mentioned that the results of Gödel and Cohen were that there were two different mathematical structures. In one, the axioms were ZFC and the continuum hypothesis. In the other, the axioms were ZFC and the negation of the continuum hypothesis. And these two structures gave two different resolutions to the uniqueness of norm problem.

David Hilbert died in 1943, three years after Gödel showed that the negation of the continuum hypothesis could not be proved in ZFC. Hilbert undoubtedly was aware of this result, and he might have regarded it as evidence (but not proof) that the continuum hypothesis could be proved in ZFC. But I'm pretty sure he would have been blown away by Cohen's result. Remember, Hilbert believed that all mathematical problems were eventually solvable, and the combined results of Gödel and Cohen showed that the continuum hypothesis, the first and arguably most important problem on his list of Twenty-Three Problems, was simply not a problem capable of solution.

I know that I was blown away when I found out that the uniqueness of norm problem had no solution; it depended on whether you wanted to work in either a structure that incorporated the continuum hypothesis as an axiom or a structure that incorporated the negation of the continuum hypothesis as an axiom.

I certainly don't speak for all mathematicians, but I know that when I am able to prove something, I feel like I've been able to establish an absolute truth. That's part of the pleasure of doing math. But there was no absolute truth involved with the uniqueness of norm problem, and at any rate, I was *way* behind the curve with regard to developments in this field. I didn't know any mathematical logic, and it appeared that if I wanted to continue to work in that area, I'd have to learn a lot of stuff that was philosophically distasteful to me. I like the bedrock certainty of establishing an absolute truth, and so establishing contrasting truths in different axiomatic systems just didn't appeal to me. That's the type of stuff logicians did, not mathematicians.

I had spent 15 years going down the road my thesis adviser had opened for me. But if I wanted to continue doing math, not just teaching it, I needed to find another road.

An Unexpected Consequence of Gödel's First Incompleteness Theorem

It is possible to understand math or science at many different levels. Probably, that's true of many other disciplines as well. And I understand what Gödel's First Incompleteness Theorem states: If the Peano Postulates for arithmetic are consistent, then there must exist theorems which are true but cannot be proved.

I don't know when the connection between this result and the supernatural (yes, that weird stuff) first occurred to me, but I felt it was sufficiently interesting that I wrote an entire book about it. And a well-known critic read it and said that he had never seen the fundamental thesis of that book before, but it made perfect sense.

So, here's the idea. In order to obtain the Peano Postulates, you've got to have the integers. If this were to exist in the Universe, we'd need something to be infinite. Maybe the Universe is infinite in extent or infinite in time, among other possibilities. These are considered to be reasonable alternatives by a significant number of cosmologists. Yes, our own neighborhood, commonly referred to as the "Hubble Bubble", started from a Big Bang and is finite in extent, but it may be only one bubble in an infinite Universe.

However, if the Universe is infinite in some sense, then there must be some theorem concerning those infinite quantities which is true but not provable. And that means there is some physical law which is true but science cannot establish it. Further, in the conventional understanding of the word, such a "law", which is true but science could not establish, would be classed as supernatural.

So imagine my surprise – and delight – when I picked up my copy of *Scientific American* in 2018 to discover that mathematicians and physicists had combined to discover the existence of an unprovable physical law. And if this one physical law is unprovable, others undoubtedly can be as well.

The brilliant physicist Werner Heisenberg said it best: Not only is the Universe stranger than we imagine, it is stranger than we can imagine. How incredibly wonderful and beautiful is it that we live in a Universe that not only has laws that we can decipher and utilize, but it also has laws about which we will remain forever in the dark. And that's the Universe that this particular mathematical quest has enabled us to see.

Chapter 6

The Bridges of Königsberg and the Four-Color Theorem

It's probably not surprising that the vast majority of mathematical quests are inaugurated by mathematicians, or at least by people who spend a lot of time thinking about mathematical questions (such as scientists and engineers). Things that happen in the real world are much more likely to kick off a scientific investigation, such as when a falling apple triggered Isaac Newton's investigation of gravity. We know that this story may be apocryphal, but nonetheless, it's a good example.

Let's face it, only mathematicians are likely to spend sleepless nights worrying about the roots of polynomials, or whether it's possible to find Pythagorean triples of integers corresponding to larger exponents than 2, as Fermat did. But this chapter is different, as the two investigations – and the branch of mathematics that connects them – were started by questions arising in the real world.

The Bridges of Königsberg

Almost every mathematician, and a lot of people who aren't mathematicians, are familiar with a classic problem that arose in the city of Königsberg sometime in the eighteenth century. The Pregel River flowed through the town, and in the middle of the river were two islands. Bridges had been built

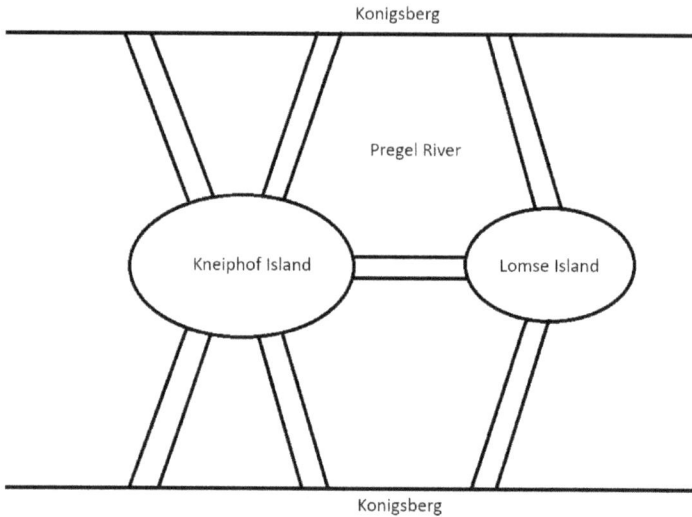

Fig. 6.1.

connecting the islands to the city; the layout looked like the one shown in Fig. 6.1.

The citizens of Königsberg enjoyed taking walks to and from the islands and wondered whether it was possible to design a path which would cross each bridge once and only once. This puzzle was resolved by Leonhard Euler in 1736; in so doing, he kicked off the branch of mathematics known as graph theory.

Euler, who would certainly be in the conversation if mathematics had a GOAT, came up with an easy-to-understand argument as to why this was not possible. Note that each of the four land masses – the two sides of the river and the two islands – each has an odd number of bridges connected to it: Kneiphof Island has five, and all the others have three. Let's say you were leaving from the near side of Königsberg (the one at the bottom of the picture), and after you finished walking over all bridges, once and once only you finished at Kneiphof Island. You would have had to both enter and leave Lomse Island. But entering *and* leaving Lomse requires two different bridges, so you would have walked over an even number of bridges to enter and leave Lomse Island, no matter how many times you did it. And no matter where you start or where you end up, you run into the same difficulty.

Graph Theory

So, what is graph theory?

Euler did what a lot of mathematics does: Simplify real-world situations in the hope of making them easier to understand. He replaced each land mass in Fig. 6.1 by a dot, and each bridge by a line. The dots are now called vertices, and the lines are called edges. The degree of a vertex is the number of edges connected to it.

There is an easy generalization of Euler's observation about the bridges of Königsberg. If a graph has more than two vertices and every vertex is of odd degree, then it is impossible to traverse each edge just once. The argument is essentially the same as the one Euler used for the bridges of Königsberg. Suppose that you start at vertex A and end up at vertex B, having traversed all the edges just once. Since there are more than two vertices, there is a vertex C. You're not there now, and you didn't start there, so each time you entered vertex C, you also left it, making an even number of total edges connected to C. Since C has an odd number of edges connected to it, the assumption that we traversed every edge exactly once must be wrong.

Note that we do need that third vertex. If the graph only has two vertices, we can connect them by any number of edges and just traverse each edge once. Although Euler did come up with a number of rules for graph theory based on the Bridges of Königsberg problem, he was, speaking mathematically, a pretty busy guy, and there were many other areas of mathematics to investigate.

But in the nineteenth century, someone asked a question for which graph theory was a natural fit.

The Four-Color Theorem

Cartography, the art and science of making maps, has been around for millennia. But the process of making colored maps, in which different countries and states were printed in different colors, started sometime in the 1850s. And it was in October of 1852 that Francis Guthrie, a student in London, England, was coloring a map and observed something curious: He needed a minimum of four colors to ensure that two countries which shared a common boundary would have different colors. After all, if you were to color France and Germany, which shared a common boundary, the same color and looked at the map from a distance, how could you tell where France stopped and Germany began?

The fact that you need a minimum of four colors can be easily demonstrated by visualizing a pizza with a surrounding crust. Cut the pizza (the part inside the crust) into three different circular sectors, and imagine that the crust is one country, and each of the three portions of each sector that don't include the crust (that's where the mushrooms and sausage go) is another country, resulting in a total of four different countries. You need three different colors; we'll use red, yellow, and blue to color the pizza slices because each slice borders the other two. But now, you need a fourth color to color the crust because the crust borders each of the red, yellow, and blue sectors.

Guthrie wondered whether any map could be colored using just four colors. Recognizing that this was a problem which could be mathematically formulated, he wrote his younger brother Frederick, who was a student at University College. After a few failed attempts, Frederick communicated the problem to his instructor, Augustus De Morgan, a well-known mathematician. De Morgan also could not solve it and forwarded the problem to a *really* well-known mathematician, Sir William Rowan Hamilton.

So, let's see how the map problem becomes a problem of graph theory. Simply imagine that the capital of every country is a vertex, and we construct an edge between two vertices (capitals) if the countries corresponding to the two countries share a common boundary. Having done that, the problem that Guthrie asked becomes this: Can we use a palette consisting of four colors and color the vertices in such a way that no two vertices which are joined by an edge have the same color?

There now occurred a lull in work on the problem lasting a quarter of a century. Usually, this occurs in mathematics when a major development of some sort, having carved out a road which most of the investigators are pursuing, hits a roadblock. That doesn't seem to be the case with the four color problem; it just seems that everybody was too busy doing other things to bother to look at it. However, in 1878, another eminent mathematician, Arthur Cayley, asked in the mathematical section of the Royal Society notes whether anyone had yet found a solution.

A Diversion: The Cayley–Hamilton Theorem

In strict terms, this section has nothing to do with the four-color theorem, and if you're not interested in matrices you can certainly skip it. But since Cayley and Hamilton both felt that the four-color problem was worth investigating, it's worth looking at a theorem that forever links them: the Cayley–Hamilton Theorem.

In order to appreciate the Cayley–Hamilton Theorem, you have to know how to multiply matrices and compute determinants, so if you don't, feel free to skip this section. And I'm not going to prove the theorem in its full generality; I'm just going to give the proof for 2×2 matrices. If $\begin{bmatrix} a & b \\ c & d \end{bmatrix}$ is a 2×2 matrix, its characteristic equation is constructed by setting equal to 0 the polynomial in the variable λ formed by taking the determinant of the matrix $\begin{bmatrix} a - \lambda & b \\ c & d - \lambda \end{bmatrix}$. The Cayley–Hamilton Theorem states that a matrix satisfies its characteristic equation if we substitute the original matrix for λ in the above characteristic equation.

The characteristic equation is the determinant of $\begin{bmatrix} a - \lambda & b \\ c & d - \lambda \end{bmatrix}$, which is $0 = (a - \lambda)(d - \lambda) - bc = \lambda^2 - (a + d)\lambda + ad - bc$; the only change that needs to be made to this scalar equation to interpret it as as a matrix equation is to think of the scalar $ad - bc$ as the matrix $(ad - bc)I = \begin{bmatrix} ad - bc & 0 \\ 0 & ad - bc \end{bmatrix}$ (where I is the 2×2 identity matrix).

We can now prove the Cayley–Hamilton Theorem for 2×2 matrices. Just computing, we see that if we substitute $\begin{bmatrix} a & b \\ c & d \end{bmatrix}$ for λ and $\begin{bmatrix} ad - bc & 0 \\ 0 & ad - bc \end{bmatrix}$ for $ad - bc$, we get

$$\lambda^2 - (a + d)\lambda - bc$$

$$= \begin{bmatrix} a & b \\ c & d \end{bmatrix} \begin{bmatrix} a & b \\ c & d \end{bmatrix} - (a + d) \begin{bmatrix} a & b \\ c & d \end{bmatrix} \mp \begin{bmatrix} ad - bc & 0 \\ 0 & ad - bc \end{bmatrix}$$

$$= \begin{bmatrix} a^2 + bc & ab + bd \\ ac + cd & bc + d^2 \end{bmatrix} - \begin{bmatrix} a^2 + bd & ab + bd \\ ac + cd & ad + d^2 \end{bmatrix}$$

$$+ \begin{bmatrix} ad - bc & 0 \\ 0 & ad - bc \end{bmatrix} = \begin{bmatrix} 0 & 0 \\ 0 & 0 \end{bmatrix}.$$

The Cayley–Hamilton Theorem is actually true for any $n \times n$ matrix, not just the 2×2 case, but the proof of the general case goes a lot deeper into linear algebra. Nonetheless, it's a lovely theorem.

Meanwhile, Back at the Ranch

Maybe it was the combination of Hamilton and Cayley that stimulated interest in the four-color problem, for within about a year there were two purported proofs. One was by Alfred Kempe, who was not only a mathematician who was good enough to be elected to the Royal Society but was also a successful barrister.

At any rate, Cayley's note stimulated Kempe to look at the four-color problem, and a year later he believed he had a proof. When he submitted his proof, he included the following note to the editor of the journal to which he was submitting it:

> *Some inkling of the nature of the difficulty of the question, unless its weak point be discovered and attacked, may be derived from the fact that a very small alteration in one part of a map may render it necessary to recolour it throughout. After a somewhat arduous search, I have succeeded, suddenly, as may be expected, in hitting upon the weak point, which proved an easy one to attack.* [1]

The proof Kempe submitted was extremely ingenious, and although his career in mathematics was long and distinguished, it is this proof for which he is best known, despite the fact that it was later shown to be incorrect. Kempe's proof was accepted by the mathematical community for 11 years, but then Percy Heawood discovered an extremely subtle error. Even though Kempe's proof was found to be flawed, there were some extremely original and very useful ideas contained in it. It was possible to modify his proof to show that five colors were sufficient to color any map in the plane, but that was sort of a consolation prize.

The fact that five colors were sufficient to color any map in the plane enabled mathematicians to redirect their efforts. It is obvious that any map

---

containing only four countries can be colored using four colors, although four colors may be needed, as the earlier pizza example shows. So suppose the four-color theorem is false. Then, there must exist a map which requires five colors to color it. In fact, there is some number N such that there is a map of N countries requiring five colors to color it, but all maps of $N - 1$ countries can be colored using only four colors.

This is a very standard approach in mathematical problems that involve positive integers. Either all integers are "good" or there is a smallest "bad" integer. If there is a smallest "bad" integer, which in this case would be the smallest number of countries that would be in a map requiring five colors (or the equivalent graph theory formulation), maybe something can be done to show that this somehow leads to a contradiction.

And that was the road that mathematicians opted to take.

The Quest Turns Ugly

Math and science have a lot in common. They are both searches for the truth with an aesthetic that prefers beauty to ugliness. The brilliant physicist Richard Feynman put these into perspective. For nearly a century, physics has been searching for a unified field theory – a simple explanation linking the four known forces in the physical world: gravity, electromagnetism, and the strong and weak nuclear forces. Feynman said that it would be great if there were such a theory, but if reality were like an onion and you kept peeling off layers only to discover a deeper layer underneath, well, that's the way it is.

Kempe's purported proof was very beautiful, but unfortunately it was flawed. Mathematicians could see no way of patching it, but they could see potential light at the end of the tunnel if they went with the assumption that the four-color theorem was false, and there was a minimal configuration that required five colors.

I've opted to include the two key definitions that mathematicians decided to work with once they embarked upon this path. If you don't understand them, don't worry; I didn't either when I first read them, and I'd have to spend some time with them, and I'd probably have to look at some examples before I felt comfortable with the definitions:

(1) An *unavoidable set* is a set of configurations such that every map that satisfies some necessary conditions for being a minimal non-four-colorable triangulation (such as having a minimum degree of 5) must have at least one configuration from this set.

(2) A *reducible configuration* is an arrangement of countries that cannot occur in a minimal counterexample. If a map contains a reducible configuration, the map can be reduced to a smaller map. This smaller map has the condition that if it can be colored with four colors, this also applies to the original map. This implies that if the original map cannot be colored with four colors, the smaller map cannot either, and so the original map is not minimal.

See what I mean?

But there was a light at the end of the tunnel, although distant and far away. If one could classify all the theoretically possible reducible configurations and then show that each one led to a map which could be colored using four colors, the theorem could be proved.

But this would be a Herculean task: first to classify the reducible configurations and then to tediously work your way through each one of them. And, in the early 1970s, help arrived from an unexpected – and to this day still dubiously regarded – source.

The Appel–Haken Proof

In 1976, Kenneth Appel and Wolfgang Haken managed to show that there was an unavoidable set of reducible configurations, which showed that a minimal counterexample to the four-color theorem could not exist. To do so, they showed that there were a total of 1,936 reducible configurations, and each one of these needed to be checked individually. This was a job that was so time-consuming that there was no other option than to do it by computer. Computers back in the 1970s were nowhere near as fast as they are today, but they were still lightning fast when compared with human beings, and it took over a thousand hours of computer time to verify the 1,834 individual cases.

In effect, the proof consisted of two parts: the human part of showing that there were 1,936 reducible configurations, and the computer part of checking each one of these configurations. Although the mathematical community generally accepts the truth of the four-color theorem, there are certainly holdouts who are uncomfortable with the idea of using a computer to verify a mathematical theorem. Of course, those who are uncomfortable with the idea of using a computer to verify a mathematical theorem have no problem with using computers to find numerical approximations for problems which have no solutions; we'll see some of these later.

And the mathematical community would undoubtedly be eternally grateful if someone would show up with an indisputable proof of the four-color theorem that did not require the assistance of a computer to check it. But this particular camel now has its nose under the tent, and I'm guessing that the future will see other theorems proved with computers, maybe even some with the intrigue of the four-color theorem.

The Four-Color Theorem for Other Surfaces

In strict terms, the four-color theorem applies to maps on a plane – the type of map that you see in a book or an atlas or that you used to fold up and place in your glove compartment before the arrival of GPS navigation. But of course, there are lots of other surfaces on which we can draw maps.

It's probably not surprising that maps on a sphere can be colored using four colors because there are ways to take a map on the sphere and transform it into a map on a plane. One such way of doing this is the Mercator projection, which has been used since the Flemish cartographer Gerardus Mercator devised it nearly four centuries ago. It projects the sphere onto a cylinder and then unwraps the cylinder, giving us the familiar "north is up, south is down" that we are so accustomed to seeing in all maps.

You might think that since the plane is the simplest possible two-dimensional object and one that has been studied for more than two millennia, the problem of coloring maps on other surfaces would be even more difficult than it is on the plane. And that's where you'd be in for a surprise. For instance, if you decided to make a map on a Mobius strip, it's been known since 1910 that six colors are enough. Admittedly, the only creatures seen to date inhabiting a Mobius strip are the ants of the Dutch artist M.C. Escher – but you never know.

And there are lots of toroidal – doughnut-shaped – worlds in science fiction, and any map on them can be colored using seven colors – or less. So, it isn't necessarily true that the simpler the object, the simpler the properties of the object – at least, from a mathematical standpoint.

So, as of now, the book on the four-color theorem is closed. But if you should come up with a straightforward proof, one that doesn't involve unavoidable sets and reducible configurations, the mathematical community would be happy to hear from you.

Chapter 7

The Quests of the Twentieth Century

It's worth getting into our time machine and going back to the dawn of the twentieth century, when David Hilbert set out his Twenty-Three Problems. Each was deemed by Hilbert to be of great importance – and great importance has certainly been the starting point for many a quest. But before taking a look at Hilbert's problems, it's worth taking a short look at David Hilbert. After all, they were his problems, and so respected was Hilbert by the mathematical community that these problems did indeed serve as guideposts.

Hilbert was, as you might suspect, an immensely talented mathematician. He made substantial contributions to many different mathematical fields, including algebra, geometry, topology, functional analysis, and differential equations. But he was almost as talented a physicist, even though until 1912 he was exclusively a mathematician who was interested more in the abstract branches of the subject than the applied ones. But in 1912, he turned to the study of physics, even to the extent of hiring a physics tutor.

In 1905, Albert Einstein had come up with the special theory of relativity, but that theory was incomplete, as it did not include gravity. Einstein spent the next decade working out how to fit gravity into the relativity framework. In the summer of 1915, Hilbert invited Einstein to join him in Göttingen, where Hilbert was the director of the Mathematical Institute. In the fall of 1915, Einstein published several papers which incorporated gravity and described the general theory of relativity. At virtually the same time, Hilbert published his treatment of an identical theory. Hilbert gave full credit to Einstein as the originator of the theory, and there was never a dispute between them as to who was responsible for the general theory. And, oh, yes, Hilbert also made several substantial contributions to the emerging theory of quantum mechanics, including the mathematical contribution known as Hilbert space, the mathematical framework upon which much of quantum mechanics rests.

But back to the Twenty-Three Problems. In reading them, one cannot help but be impressed by the scope and range of Hilbert's familiarity with many fields of mathematics. Frankly, even understanding several of the problems is beyond my capability; I can read the problems and appreciate approximately where they lie, but it would require some study before I could say I understood what each problem seeks to resolve.

There is, however, one important gap that I observed when reading the problems: There is nothing concerning either probability or statistics. Hilbert certainly knew of the existence of these fields, but possibly no problem had emerged in them that Hilbert felt was sufficiently important to warrant inclusion among the Twenty-Three. One possibility is that many of the other branches of mathematics intertwine with each other to a greater extent than do probability and statistics.

Nonetheless, Hilbert's Twenty-Three Problems were such a landmark that it's worth taking a look at some of them – or at least the ones that can be understood without taking high-level math courses – a century and a quarter after they were propounded and see where we stand.

Hilbert's First Problem involved the continuum hypothesis. The way that Hilbert stated the problem involved either showing that a set with a cardinal between that of the integers and that of the continuum (the closed interval [0, 1]) either existed or didn't exist. It was shown that it was impossible to resolve this question within the framework of the Zermelo–Fraenkel axioms for set theory. There currently does not exist a consensus as to whether or not this resolves Hilbert's First Problem. I would hesitate to speak for Hilbert, but at the time he stated the First Problem, I don't think he would have thought of the question as existing outside of the Zermelo–Fraenkel axioms. At the time, people just weren't looking at other axiomatic systems.

Hilbert's Second Problem was to show that the axioms of arithmetic were consistent. Gödel was able to show that it was impossible to prove that the axioms of arithmetic were consistent within the framework of arithmetic itself. Again, there is no consensus as to whether this resolves Hilbert's Second Problem, but I'm on somewhat of the same side that I was in the previous paragraph. It isn't hard to construct an inconsistent set of axioms containing the axioms of arithmetic – simply stick in a ridiculous axiom. That would not invalidate the axioms of arithmetic, on the same line of thinking that if you have a sealed package of apples within an apple barrel and one of the apples outside the sealed package is rotten, you're probably still going to be willing to eat the apples from the sealed package.

At any rate, that brings us to Hilbert's Third Problem, and an escape from the mind-numbing (at least, for me) question of consistency of axiom sets.

Dissections and Hilbert's Third Problem

I've always been a science junkie, but I skipped taking biology during my sophomore year in high school because I knew you had to dissect a frog and maybe a cow's eye. Talk about yucky! But dissections in the mathematical world aren't yucky at all.

A simple example of a dissection is seen in Fig. 7.1, which represents a proof of the equality $a^2 - b^2 = (a-b)(a+b)$. Of course, we can verify this by simply expanding the right-hand side, but that wouldn't be a dissection. If we simply cut the figure for $a^2 - b^2$ along the dotted line and reposition the lower rectangle adjacent to the upper rectangle so that the two sides

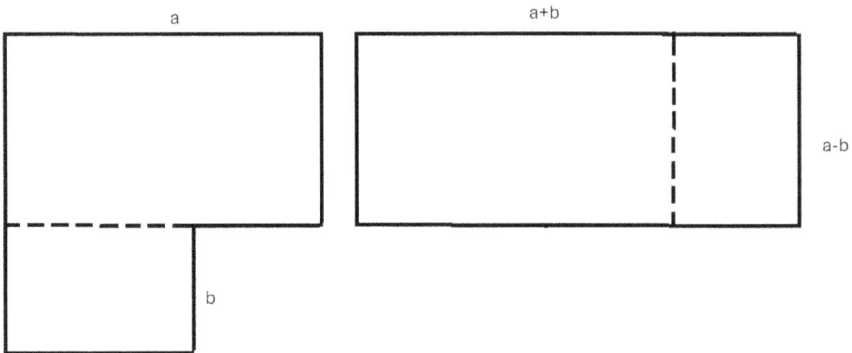

Fig. 7.1.

that are $a - b$ in length touch each other as in the diagram, we can see that $a^2 - b^2$, which is the area of the larger square with the smaller one removed, is the area of a rectangle whose two sides are $a - b$ and $a + b$.

It's also possible to do dissections in three dimensions. In fact, you do it all the time when you slice something, be it a loaf of bread, a cake, or a roast chicken. Hilbert's Third Problem is simple to state: Given two polyhedra of equal volume, is it possible to dissect the first one into a finite number of polyhedral slices (which means that all the cuts need to be straight) and reassemble those pieces into the second?

This was the first of the Hilbert Twenty-Three to be resolved, and perhaps it was fitting that Max Dehn, who was one of Hilbert's students, was the individual to do so. Dehn showed that this was not possible. Although Dehn's proof is beyond the scope of this book, it is possible to give an indication of the key idea in the proof – the use of an invariant quantity.

There are a lot of words in the mathematical vocabulary that are not descriptive. A field in the mathematical sense bears no resemblance to any field found in the real world, and the word "normal" is used in many different areas, and almost never describes a normal situation. For example, the normal to a plane is a line perpendicular to it, and if you were to just choose a random line, it is almost certain that line would not be perpendicular to the plane. But the word "invariant" is just that – a quantity which does not change under certain processes.

There are several nice examples of invariant quantities associated with the problem of covering an 8×8 checkerboard with tiles that look like dominos – a 1×2 rectangle, as shown in Fig. 7.2.

Granted, this probably wouldn't be much of a challenge even for a five-year-old, as long as there were enough dominoes. But what about if we removed one of the corner squares, as shown in Fig. 7.3?

Again, the five-year-old will probably see fairly quickly that you can't do this, and the reason why this cannot be done can be explained in terms of invariants. We start out with an odd number of uncovered squares, and adding a tile still leaves an odd number of uncovered squares. The fact that the number of uncovered squares is odd is invariant under the covering process, and when we finally get down to the end, you can't cover a single

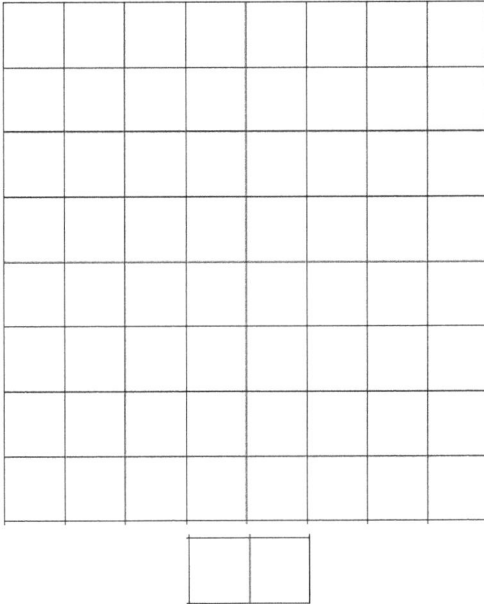

Fig. 7.2.

Fig. 7.3.

Fig. 7.4.

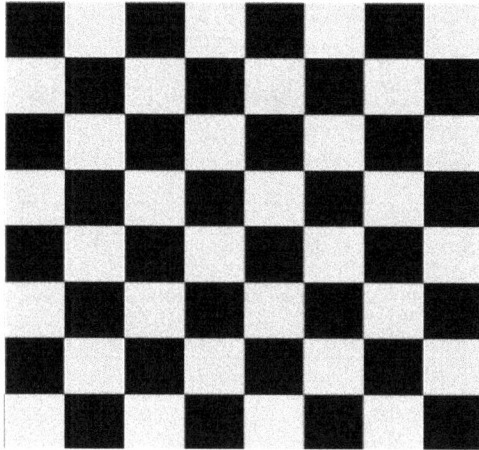

Fig. 7.5.

uncovered square with one of the dominos without having the last domino extend beyond the checkerboard with one square removed.

But what if we removed the other corner square on the diagonal, leaving the checkerboard as shown in Fig. 7.4?

This is a *much* harder problem because the key invariant is hidden. That key invariant can be seen if we look at the original checkerboard (Fig. 7.5).

When we remove the upper-left and lower-right corner squares, we are left with a board that has two more white squares than it does black squares. The key invariant here is the number of uncovered white squares minus the number of uncovered black squares. This doesn't change when we cover two squares with a domino, and it is always equal to 2. So, when there are two squares remaining, the invariant will still be equal to 2, and thus the last two uncovered squares will both be white.

And that, in principle, is what Max Dehn did to solve Hilbert's Third Problem. He found an invariant, now known as a Dehn invariant, that was preserved under all dissections. Unlike the invariants associated with the checkerboard problems above, Dehn invariants are tensor products, which require an explanation beyond the scope of this book. But the key thing is that the cube and the regular tetrahedron (a four-sided pyramid all of whose faces are equilateral triangles) have different Dehn invariants, and so you can't slice a tetrahedron into a finite number of pieces and reassemble it as a cube even if their volumes are the same.

I'm not sure whether Hilbert had a rooting interest in which way the solution to his Third Problem would go (could you do it or not), but I'm sure he was proud of the fact that the first of his problems to be solved was solved by one of his own students.

Hilbert's Seventh Problem

Pythagoras is famous not only for the Pythagorean Theorem but also for having said, "All is number". This statement still resonates more than two millennia later, as we wrestle with the extent to which the Universe can be digitized. But when Pythagoras uttered those words, even though the Greeks knew of numbers such as the square root of 2, it was felt that all numbers were rational – either integers or ratios of integers such as the fraction 3/5.

At least, that's how they felt until Hippasus of Metapontum showed that the square root of 2 could not be expressed as a ratio of two integers. This knowledge was believed to be so dangerous to Greek society that an effort was made to keep it secret from the general public, and Hippasus himself disappeared overboard while on a sea voyage. Conspiracy theorists then and now have had a field day with that.

But mathematicians – and the general public – eventually came to accept the existence of algebraic numbers, such as the square root of 2, which were the roots of polynomials with integer coefficients. Of course, there were other well-known numbers such as π and e, which could not immediately be shown to fit into that framework.

At the time Hilbert attended college, it was well known that the algebraic numbers were countable, that is, they could be put in one-to-one correspondence with the integers. The proof of this relies on the fact that a countable union of at-most-countable (finite or countable) sets is countable. This isn't too difficult to establish, that is, if you are Georg Cantor. The proof goes like this. Assume that each of the countable sets is labeled $A_n = \{a_{nk}: k = 1, 2, \ldots\}$; if the set is finite, we can simply write $a_{nk} = a_{nN}$ for $k \geq N$, where the first N elements of A_n are distinct.

We now write the union of $\{A_n: n = 1, 2, \ldots\}$ as a list as follows: a_{11}; a_{12}, a_{21}; $a_{13}, a_{22}, a_{31}, \ldots$, where we first write down all the a_{jk} whose subscripts total 2, then the ones whose subscripts total 3, then the ones whose subscripts total 4, and so on. If we come to an element that already appears on the list, we simply don't bother to include it.

We can use the result of this proof over and over again to demonstrate the countability of the algebraic numbers. Our first task is to show there are only a countable number of polynomials with integer coefficients. First of all, there are only a countable number of polynomials of degree 1 because there are only a countable number of such polynomials whose coefficient of x is 1, and there are only a countable number of such polynomials whose coefficient of s is 2, and so on. Since the union of a countable collection of countable sets is countable, there are only a countable number of polynomials of degree 1.

We now use induction. Assume that we have established that there are only a countable number of polynomials of degree n. So, there are only a countable number of polynomials of degree $n + 1$ whose leading coefficient is 1, only a countable number of polynomials of degree $n + 1$ whose leading coefficient is 2, and so on. The union of this collection of polynomials is therefore a countable union of countable sets and hence countable. Next, the set of all polynomials with integer coefficients is the countable union of all polynomials of degree n with integer coefficients, so the set of all polynomials with integer coefficients is countable. And finally, each polynomial has a finite number of roots – the algebraic numbers – and so, applying our (by now overworked) theorem again, the set of all algebraic numbers is countable.

It was also known that the real numbers were uncountable, and so the transcendentals (I don't know why they're called that), which are the real numbers that aren't algebraic, are also uncountable. It was suspected that π and e – and a host of other well-known numbers – were transcendental. But it was Ferdinand von Lindemann, the mathematician who supervised Hilbert's doctoral dissertation, who finally proved that π was a transcendental number – a number which could not be the root of a polynomial with integer coefficients. The reaction of the mathematical community – and the world at large – to Lindemann's result was considerably warmer than had been the reaction to Hippasus' result almost 2,500 years earlier. At the very least, there is no indication that Lindemann was hesitant to embark on voyages, by sea or otherwise.

The imprint of Lindemann's result can be seen in Hilbert's Seventh Problem, which is to prove that any number of the form a^b is transcendental, where a is an algebraic number unequal to either 0 or 1 and b is an irrational algebraic number, such as the square root of 2. This result was established independently in 1934 by Aleksandr Gelfond and Theodor Schneider.

The Kepler Conjecture

We live in what might be called the age of Big Data, Go back four and a half centuries, though, and you'll find what was probably the first example of Big Data – the observations compiled at Uraniborg, undoubtedly the first modern scientific establishment. Built by the Danish astronomer Tycho Brahe, its observations were the foundation for one of the most important pieces of data fitting, the Three Laws of Johannes Kepler.

Like many modern scientists confronted with a batch of data, Kepler had a theory. At that time, there were five known planets in addition to Earth: Mercury, Venus, Mars, Jupiter, and Saturn. As a mathematician, Kepler knew there were five regular solids: the tetrahedron, cube, octahedron, dodecahedron, and icosahedron. Kepler believed that the orbits of the planets were somehow related to the properties of the regular polyhedra, and so for years he tried to hammer the square peg of Brahe's data into the round hole of this theory. After more than 40 failed attempts, he decided to trust the ugly data rather than the beautiful theory. Instead, he made the assumption that the orbits of the planets were in the shape of an ellipse, with the Sun at one of the foci. The data fitted this theory beautifully.

Kepler was able to do more than just fit the data to an ellipse. His Three Laws of Planetary Motion would provide some of the validation for Isaac Newton's Theory of Gravitation, as those Three Laws could be proved mathematically from Newton's model for the gravitational force. And Kepler managed to come up with this at the same time that he had other, possibly more important, matters to attend to. Kepler successfully defended his mother against a charge of witchcraft. Considering the beliefs of that era, that might have been an even bigger accomplishment than the creation of the Three Laws of Planetary Motion.

But Kepler was not just a data-fitter; he was a mathematician of some accomplishment. One of the problems that attracted his interest was the problem of sphere packing. Given a collection of spheres, what is the most effective way of packing them so that they fill the greatest volume of the available space? Kepler conjectured that the best way to do this was by the method used to stack cannonballs. While not easy to describe mathematically, you see an example of this whenever you go to a grocery store and see a pyramidal array of grapefruit stacked on top of each other. Each layer of grapefruit is packed regularly, with the bottoms of one layer fitting into spaces of the layer directly below it. This method of packing fills about 75% of the available space.

The proof of this conjecture finally arrived in 1998 in a computer-assisted proof by Samuel Ferguson and Thomas Hales. Acceptance of the proof took almost a couple of decades, but it was finally accepted in 2014. Recently, Hales published the story of this quest; it is freely available as an arXiv publication (https://arXiv.org/html/2402.08032v1). It traces a short history of unsuccessful attempts (including one by the architect Buckminster Fuller) and is mostly in English. It is readable if you're willing to skip over the mathematical shorthand (such as "branch and bound") to get the flavor of what it's like to embark upon such a quest and the frustrations one can encounter.

Chapter 8

The Quests of the Twenty-First Century

Let's begin by looking at a mathematical quest that actually made headline news within the past decade.

Twin Primes Conjecture

As has been mentioned, prime numbers have been known to play an important role in mathematics since the ancient Greek mathematicians. We've seen that Euclid was able to prove more than two thousand years ago that there are an infinite number of primes. There are a lot of primes among the small numbers – 2, 3, 5, 7, 11, 13, 17, and 19 are the primes less than 20. That's 40% of all numbers less than 20.

Let's recap a little by way of introducing this topic. As you go further out, the primes get scarcer. There are 25 prime numbers less than 100, 168 less than 1,000, and 1,229 less than 10,000. Tables of prime numbers have been studied for some time, and in 1798, the French mathematician Adrien-Marie Legendre proposed that the number of primes less than a given number x could be approximated by the fraction $x / \ln(x)$, and as x becomes larger and larger, this fraction becomes an ever more accurate approximation to the number of primes less than x.

This was an insightful conjecture and one that was eventually proved to be correct, but it took almost a century to prove what has come to be known as the prime number theorem. It was eventually proved at the end of the nineteenth century by two French mathematicians, Jacques Hadamard and

Charles Jean de la Vallée Poussin. They each came up independently with a proof using methods that are substantially beyond this book. Although there have since been other proofs, none are simple.

Another way to look at the prime number theorem is that it tells us that the average distance between successive primes is approximately the natural logarithm of the first of the two successive primes – or the second, because the prime number theorem itself talks only about approximate values, not exact ones. The natural logarithm of x is a slowly increasing function, but it keeps getting larger and larger. The prime number theorem tells us, for instance, that there will be two primes separated by at least 10^{100} (this number is known as a googol, after which Google is named) if you just go out far enough in the number line.

But there are successive primes that are separated by the smallest possible distance (for primes): 2. Other than 2 and 3, successive primes cannot be separated by just 1, for if x is larger than 2, one of the pair x and $x + 1$ is an even number and so cannot be prime. Successive primes that are separated by 2, such as 11 and 13 or 17 and 19, are known as twin primes.

The twin primes conjecture states that there are infinitely many pairs of twin primes. And while it's easy to see that there are a few early on, the prime number theorem tells us that they will be increasingly rare because, as we go further out in the number line, the average distance between successive primes – the natural logarithm of the first of successive primes – becomes larger and larger.

Every so often, someone comes out of nowhere to achieve an astonishingly unexpected feat. This seems to happen with some frequency in sports. In tennis, for instance, only a few people had ever heard of Emma Raducanu prior to the United States Open in 2021 – which she won, to the amazement of practically everyone. But eight years earlier, something even more astonishing happened in the world of mathematics. Yitan Zhang, a mathematician who was probably even less well known to the mathematical community than Emma Raducanu was to the tennis community, proved an astounding and totally unexpected result.

Yitan Zhang's story is perhaps no less astonishing than the result he proved. The 1980s and 1990s saw the job market for academic mathematicians take a serious hit, which has continued to this day. Virtually all job markets are a matter of supply and demand. I was fortunate to arrive on the scene when the demand for mathematicians considerably exceeded the supply, but Zhang was not so lucky. In addition to arriving at the job market at a time when the supply exceeded the demand, Zhang and his thesis adviser at Purdue University did not get along. Even though Zhang eventually received his Ph.D., in an interview later his adviser is reported to have said that Zhang's work on something called the Jacobian Conjecture (I have no idea what that is) wasted seven years of both Zhang's and his time.

As a result, Zhang was compelled to support himself with a variety of jobs and, at one point, was living out of his car. At the time of his discovery concerning the twin primes conjecture, Zhang was a part-time lecturer at New Hampshire University, working to supplement his income by managing a Subway franchise. But this left him time to think about the twin primes conjecture, and although he was not able to prove it, he was able to make the first significant dent in the problem by showing that there were an infinite number of successive prime pairs separated by a distance of less than 70,000.

Showing that there are an infinite number of successive prime pairs separated by less than 70,000 may not seem impressive, considering that the twin primes conjecture states that there are an infinite number of successive prime pairs separated by 2. But it's a start – and for comparison, recall that the first dent in the Goldbach Conjecture was to show that every even number was the sum of not more than 300,000 primes.

As soon as the mathematical community saw – and vetted – Zhang's paper on the twin primes conjecture, that segment of the community interested in the problem began to use Zhang's ideas. Two world-class mathematicians, Terence Tao and James Maynard – both Fields Medalists – were able to extend Zhang's work so that the number 70,000 has now been improved to 246, where it currently stands. In other words, there are an infinite number of successive prime pairs such that the distance between them is less than 246. Stay tuned!

And what of Zhang? Several prestigious awards were followed by offers of prestigious positions. Zhang is now a tenured full professor in the Mathematics Department of the University of California at Santa Barbara. The days of having to support himself by working as the manager of a Subway franchise are now behind him.

Child's Play

I hope I've managed to convince you that mathematics involves not only the search for truth and beauty but also the search for intellectual entertainment. If you're not yet convinced, let me present two puzzles that have not yet been resolved. Both are problems a child can understand, and one is a problem that your five-year-old daughter – assuming you have one – may be able to solve, even though to date it has eluded the efforts of the world's best mathematicians!

For the first of these two puzzles, all you need to know is simple arithmetic: how to multiply, add, and divide. You don't even have to know how to subtract! For the second, all you need is a set of seven very specific pieces of cardboard (or any other stiff material, such as plastic or wood) because all that you have to do is move them around in an attempt to accomplish a particular goal.

The Collatz Conjecture

Lothar Collatz was a respected twentieth-century German mathematician who made important contributions to several fields of mathematics. However, he will probably best be remembered for what is now known as the Collatz Conjecture. It is a conjecture that requires only addition, multiplication, and division to understand. The Collatz Conjecture was regarded as so fiendishly difficult that Paul Erdös once said of it that mathematics was not yet ready to tackle such a problem.

Start with any positive integer: Let's say we decide to start with 19. If it is an odd number, triple it and add 1. If it is an even number, divide it by two. At any stage of the process, follow the same rules, depending on whether you have an odd or an even number. Note that when you perform this process on an odd number, the next number will be even, since tripling an odd number results in an odd number, and adding 1 to that gives you an even number.

Anyway, let's start with 19:

19, 58, 29, 88, 44, 22, 11, 34, 17, 52, 26, 13, 40, 20, 10, 5, 16, 8, 4, 2, 1.

At this point, were we to continue, we'd go into a repeating loop: 1, 4, 2, 1, 4, 2, 1, 4, 2, 1, ... The Collatz Conjecture states that no matter what number you start with, you'll *always* end up back at 1! So simple to state, yet, as Erdös remarked, fiendishly difficult to prove. We've barely managed to make a dent in this problem.

In this era of high-speed computers, people have tried to see what happens with various starting numbers. As of a couple of years ago, a computer demonstrated that the Collatz Conjecture holds: You always get back to 1 if you start with any number less than one quintillion. Now, one quintillion is a pretty big number in the everyday world, but just as there are an infinite number of whole numbers between 1 and infinity, there are an infinite number of whole numbers between one quintillion and infinity.

But fiendishly difficult problems are alluring for two reasons. The first is that they're fiendishly difficult. If you're a mountain climber, you would want to be the first to climb Mount Everest, and if you're a mathematician, you'd love to be able to solve a fiendishly difficult problem.

If one thinks of the Collatz Conjecture as a mountain, mathematicians planted a flag at the foothills sometime in the 1970s, when they showed that almost all sequences eventually reach a smaller number than the number from which they started. The difficulty here is that "almost all" does not mean "except for a few". As a result, we can't complete the proof by saying, "Start at N_1, then when you reach a number N_2 less than N_1, start again at N_2, then you'll reach a number N_3 less than N_2, etc. Keep going until one of these numbers is 1". The difficulty with this "proof" is that the number N_2 that you reached might be one of the exceptional numbers to which the theorem does not apply. But at least it's a start.

One of the people who have decided to take a shot at the Collatz Conjecture is Terence Tao, who helped lower Zhang's gap of 70,000 to 246. However, in order to avoid getting mired in mathematical quicksand, he limited the amount of time he would spend on the problem. During the past decade, he's gone back to the Collatz Conjecture several times. Sometimes he got nowhere, but often just looking at a problem intensely gives you a better feel for it. Additionally, most of us have experienced a situation in which we were stuck on some sort of problem – not necessarily mathematical – put it aside for a while, and then, all of a sudden, inspiration struck.

That happened to Tao with the Collatz Conjecture. He managed to improve the earlier result of the 1970s that almost all numbers will eventually hit a smaller number. That result is extremely vague, but Tao managed to show a result that looks something like this: 99% of all numbers larger than 10^{100}, when used as starting numbers for the Collatz Conjecture, eventually end up with a number less than 10^{10}. Here, the numbers 99%, 10^{100}, and 10^{10} are standing in for numbers that Tao showed must exist but whose actual values are currently unknown and indeed may never be known. But, as the intro to the TV show *The X-Files* stated, "The truth is out there". Unspoken in that intro was that we may never know that truth. And we may never know the actual values of the numbers that Tao showed must exist.

But here's what I think – for what it's worth – and that's probably not a whole lot. Unlike the Goldbach Conjecture, the Collatz Conjecture deals with three very specific numbers: 3, 1, and 2. Those are very small numbers. And computers tell us that if you start the Collatz process with a number less than one quintillion, you'll eventually hit 1. So, I'm guessing that it's true for all numbers, simply because it just seems that any process which can be simply described in terms of 3, 1, and 2 and which is known to be true for the first one quintillion numbers is probably always true.

The Seven Squares Problem

Would you believe that there is an outstanding problem that has stumped the world's best mathematicians that your five-year-old daughter might be able to solve? There is – and it's the seven squares problem. I don't know what name it goes by, but that's how I think of it.

The seven squares problem involves – not surprisingly – seven squares. One square is 201 by 201 (I've deliberately left out the units for a reason I'll mention later), and the other six are all 100 by 100. The area of the 201-by-201 square is 40,401 square units, and the area of each 100-by-100 square is 10,000 square units.

The problem is to use the six smaller squares to completely cover the larger square. Since the total area of the six smaller squares is 60,000 square units, you have more than enough material to do it, that is, if you were allowed to cut the 100-by-100 squares into smaller pieces. But you're not allowed to do so.

It is known that if you have seven 100-by-100 squares, even without cutting them up, you can cover the 201-by-201 square. But as of this writing, no one has either managed to cover the 201-by-201 square with the six 100-by-100 squares or proven that it is impossible to do so.

I suspect that most of the readers of this book are familiar with the Rubik's Cube. The 15 puzzle was the Rubik's Cube of my generation. For those not familiar with it, here's a website link: https://15puzzle.uk.

Of course, the 15 puzzle of my generation was not online. It consisted of a frame in which the tiles, either wood or plastic, could be moved either horizontally or vertically into the vacant space. Many of the puzzles that were sold came with a paper insert offering substantial prizes if you could exhibit a 15 puzzle moved into a specific configuration. I, and almost every other child that I knew, spent hours trying to maneuver the tiles into one of the prize-money configurations.

We didn't know it – although some of us, or our parents, may have suspected it – but we were "drawing dead". The puzzle was sold with 1–4 in order in the top row, 5–8 in order in the second row, 9–12 in order in the third row, and 13–15 in order in the first three positions in the bottom row, with the vacancy in the last position in the bottom row. From this position, half the possible configurations can be mathematically shown to be unreachable, and the prize-money configurations were always unreachable.

Did this frustrate us? I don't think so. Most of us who spent time with the puzzle (and that was most of us) were fascinated by it. Possibly this was the first quest on which many of us embarked. Similarly, I don't think any child who works on the seven squares problem will be harmed if someone actually manages to show that it is not possible to cover the big square with the six smaller squares. And since, as of this writing, no one knows, feel free to give it to your son or daughter and see what happens.

Now, I strongly doubt that your five-year-old daughter is capable of coming up with a proof that it is impossible to cover the 201-by-201 square with the six 100-by-100 squares. Mozart was writing symphonies when he was five, and Gauss was showing how to solve an addition problem by using multiplication when he was seven or eight. But even Gauss wasn't writing serious proofs at that age; he just saw – and used – an intriguing pattern.

You need to have squares of those exact dimensions, but in these days where we can measure stuff in nanometers, it shouldn't be difficult to man-ufacture these squares out of cardboard or plastic. Color the larger square green and the smaller ones yellow – or any contrasting colors of your choice. Describe it as a game or a puzzle to your five-year-old daughter, and let her know that if she is successful in covering the green squares with the yellow ones, she will be the first person to do so. Make sure to tell her that no one knows whether it is possible or not.

And one day, maybe you'll come home from work to find your daughter beaming proudly with the solution to a problem no one has yet managed to solve. It could happen. But more importantly, you may have planted the seed for your daughter to delight in pursuing the unknown.

The Clay Millennium Problems

January 1, 2000, ushered in not only a new century but a new millennium. An organization known as the Clay Institute decided not only to pay tribute to Hilbert's Twenty-Three Problems but also to offer monetary rewards to mathematicians who solved what it considered to be seven of the most important problems confronting mathematics in the twenty-first century. The Institute offered a prize of $1,000,000 for solutions to seven problems, henceforth known as the Millennium Prize Problems.

Two of those problems are similar to some of Hilbert's Twenty-Three Problems in that I have no idea what they're talking about, and it would require some effort for me to understand them. One of those problems – the Poincaré Conjecture – has already been solved. Three of those problems are still outstanding, but I can understand them and, hopefully, indicate in this chapter why they are so important. And one of them – P vs. NP – resulted in my only court appearance since the time when I was 17 years old and was arrested for disturbing the peace (the judge let me go with just a talking-to). That story, and the mathematics that surrounds it, seems to me to be sufficiently interesting that it warrants a chapter of its own, which will follow this one.

But first, let's take a look at the other Millennium Prize Problems – at least the ones that I can understand.

The Poincaré Conjecture

The Poincaré Conjecture deals with the following problem. Suppose you have a finite chunk of ordinary three-dimensional space, except it's made of some substance such as clay or dough, so you can shape it. You're allowed to shape it as a potter would with clay or a baker would with dough, but you are not allowed to tear or cut it. What extra condition do you need to place on it so that you could shape it into a sphere?

Well, you certainly don't want your chunk of three-dimensional space it to look like a doughnut or an inner tube; in other words, you don't want it to have any holes in it. Spheres don't have holes in them, and reshaping without tearing or cutting wouldn't get rid of the hole. So, Poincaré suggested that this restriction – that your chunk of three-dimensional space have no holes – was all you needed to shape it into a sphere.

Of course, the formal statement of the Poincaré Conjecture involves some mathematical terminology. Specifically, it states that every three-dimensional topological manifold (the chunk of three-dimensional space) which is closed, connected (not two separate chunks), and has a trivial fundamental group (no holes) is homeomorphic (can be shaped) to a three-dimensional sphere.

This problem has been around since the early twentieth century, and many of the advances made in geometric topology (basically, the study of shapes of objects) made during that century came about as a result of trying to establish the Poincaré Conjecture. But what eventually led to resolving the Poincaré Conjecture came about as the result of the American mathematician Richard Hamilton's study of the problem using partial differential equations known as Ricci Flows.

Partial differential equations are extremely important mathematical objects; we'll discuss them in a little more detail when we talk about another of the Millennium Problems – the Navier–Stokes Equation. But for now, it's enough to know that Hamilton's work on studying the Poincaré Conjecture using Ricci Flows is acknowledged to be a major step in the eventual establishment of the Poincaré Conjecture.

The actual proof of the Poincaré Conjecture was completed by the Russian mathematician Grigori Perelman in the early twenty-first century. As we have seen throughout this book, mathematics has produced its share – perhaps more than its share – of eccentrics. Perelman continues that tradition in his own way.

Perelman's exceptional talent was recognized early in his life. He did his initial work as a student at Leningrad State University, after receiving a special medal for achieving a perfect score in the International Mathematics Olympiad. Like many other disciplines, mathematics recognizes special achievements with prizes and awards. The list of awards Perelman has received is astounding, but what is even more surprising is that he has rejected most of them.

In 2006, he was awarded the Fields Medal. The Fields Medal is even more exceptional than the Nobel Prize. The Fields Medal is awarded only every four years, and if the committee responsible for awarding it doesn't feel that anything done in the past four years merits a Fields Medal, they don't award it. Perelman declined the award.

In 2010, the committee responsible for awarding one of the Clay Institute Millennium Prizes decided that Perelman's proof of the Poincaré Conjecture merited an award. Perelman declined that award, too – and the $1,000,000 prize that came with it – saying that Hamilton's work was no less important than his and that Hamilton should be recognized equally by the committee.

I don't know Grigori Perelman, but in the unlikely event that he should read this book, here's how he can cut through this particular Gordian knot. Accept the prize and give half the money to Hamilton. I can guarantee that Hamilton will thank him profusely and that the half-million dollars Perelman would end up with would go a long way to helping him support his mother and sister, who seem to be the two people with whom he spends the most time.

As far as I can ascertain, Perelman has adhered to his principles. He lived quietly with his mother in St. Petersburg and visited his sister occasionally in Stockholm. He was offered numerous full professorships with tenure at the most prestigious universities in the world, but he turned them all down.

As far as we know, one of the world's greatest mathematicians no longer does mathematics – but who knows? Maybe someday soon, a paper from Perelman will appear, showing that every even number is the sum of two primes.

The Navier–Stokes Equation

Let's face it, no matter how intriguing the Poincaré Conjecture was or how important it is considered to be by the world's great mathematicians (or at least, the world's great topologists), the resolution of the Poincaré Conjecture is likely to have no impact on your life – unless, of course, you are one of the world's great topologists. And if you are, thanks for considering this book worthy of your time.

But you're probably not one of the world's great topologists, and your life most definitely will be affected by major breakthroughs in solving the Navier–Stokes Equation. The Navier–Stokes Equation describes the motion of fluids. There are any number of fluids which impact our lives – water and air being the two most immediate that come to mind. Their motion is described in terms of waves. We are all familiar with water waves, from the smooth to the violent, and anyone who has ever experienced sudden air turbulence in an airplane is all too familiar with the realization that our fate depends on our understanding of these waves.

The Navier–Stokes Equation is a partial differential equation and is almost two centuries old. Sir George Gabriel Stokes was one of the great mathematical physicists of the nineteenth century, making important contributions in both physics – especially in hydrodynamics – and mathematics. Stokes' Theorem in multidimensional calculus is one of the most profound results in the subject and generalizes Green's Theorem for two dimensions to three (and higher) dimensions.

But what may well be of more interest is how Stokes viewed himself. Charles Darwin, in his later years, felt that he had become simply a machine drawing conclusions from large collections of data. Stokes felt the same way. He became engaged in his late 30s, and it was not one of history's most passionate romances. His fiancée sensed this and gave signs of wanting to call off the impending marriage. Stokes wrote to her, "Then it is right that you should even now draw back, nor heed though I should go to the grave a thinking machine unenlivened and uncheered and unwarmed by the happiness of domestic affection". The marriage went ahead, and though Stokes continued to make significant mathematical and scientific contributions, they did not have the same importance and impact as his pre-marital contributions.

The latter portion of Stokes' career was consumed with honorary and administrative appointments, such as being Secretary of the Royal Society. P. G. Tait, a contemporary, noted that, "What a comment on things as they are is furnished by the spectacle of genius like that of Stokes' wasted on the drudgery of Secretary to the Commissioners for the University of Cambridge; or of a Lecturer in the School of Mines; or the exhausting labor and totally inadequate remuneration of a Secretary to the Royal Society".

And what of Navier? Besides being the most likely holder of the prize for mathematician with the most first and middle names (Claude Louis Marie Henri Navier), he was mentored by Fourier, who became a lifelong friend. Fourier, of course, was one of the leading mathematicians and physicists of his day. Navier's interest was in civil engineering, and he became the leading bridge-builder and designer of that era. As a result of Fourier's influence, Navier helped transform the engineering curriculum by emphasizing not only the practical aspects but also the application of mathematics and physics to engineering problems.

Navier was recognized during his lifetime as a pre-eminent engineer, but as sometimes happens, Navier's future acclaim came from the work he did on adapting Euler's equations to describe first the flow of an incompressible fluid and then to take into account viscosity. This work led to the Navier–Stokes Equation. Two centuries later, its mysteries still remain to be unraveled.

Although a discussion of the Navier–Stokes Equation itself is not appropriate for this book, one of the questions that remain to be answered is easily understood. Under what conditions do solutions to the equation exist, and are those solutions unique? Having a unique solution to a differential equation means that only one thing can occur. Since the Navier–Stokes Equation is one of the pillars on which weather forecasting is based, if we could be certain that solutions are unique, we could be considerably more confident in our ability to predict the weather days, weeks, or months from now.

All But One of the Remaining Millennium Problems

All but one of the eight Millennium Problems require a lot of high-powered math to even understand what the problem is. The Riemann Hypothesis is actually a holdover from Hilbert's Twenty-Three Problems; it deals with the zeros of the Riemann zeta function and has important ramifications, most of which are unknown to me and are well beyond the scope of this book. The Birch and Swinnerton-Dyer Conjecture involves elliptic functions, which played a key role in Wiles' proof of Fermat's Last Theorem. The Yang–Mills Mass Gap Problem has something to do with equations from quantum mechanics. I can't even come close to understanding the Hodge Conjecture, which has to do with a branch of topology that totally confused me when I was a graduate student – and would probably confuse me even more if I looked at it now.

Incidentally, this bears upon the public perception of the knowledge of mathematicians. Over the years, I've told many people I was a mathematician – I'm not ashamed of it – and have generally received one of two reactions. The first is surprise that there's mathematics beyond subjects studied in high school, although this has been lessened thanks to numerous TV shows and movies in which mathematics or mathematicians occupy a central position. The second is that every mathematician knows all about mathematics.

Yes, there are some mathematicians whose knowledge of mathematics is encyclopedic. Paul Erdös was one; he authored or co-authored over 1,400 papers in virtually every field of mathematics. But most mathematicians are like me – we're specialists. I know most of the undergraduate mathematics curriculum, but I've really only worked on a few problems in highly specialized areas. Several of these problems achieved quest level, and when you're wrapped up in a quest, there really isn't much room for anything else. I seriously doubt if Indiana Jones cared about the Super Bowl.

But there's one final problem in the Millennium Problems that has two – maybe three – attributes that merit an entire chapter in this book. The first is that it's easily understandable. The second is that it may have the most significant impact, not only on the development of mathematics but on everyday life. And the third is that it played an unexpected role in one of the most difficult experiences I have ever encountered.

.

Chapter 9

The Traveling Salesman Problem

I've spent most of my life investigating mathematical problems. Some of those investigations have been successful, some not – and I imagine that's true of most mathematicians. But outside the world of mathematics, most of those investigations were inconsequential.

I never would have imagined that I would get involved in one of the most famous quests of twenty-first-century mathematics and that it would result in threats of violence to myself and my family so severe that it forced me to go to court to obtain a restraining order against someone who had been my former student.

Bob

I first encountered Bob (not his real name) in the 1980s as a student when I was teaching Introduction to Analysis, a standard upper-division course for math majors, at California State University, Long Beach (CSULB). Bob had graduated some years before with a degree in mathematics from one of the branches of the University of California. But he had decided to pursue a master's degree at CSULB and felt he needed a refresher before taking graduate courses. He asked if he could sit in on my lectures, to which I readily assented. This is one of the courses I most enjoyed teaching, partly because I had considerable difficulty understanding the material the first time I encountered it. As a result, I can appreciate the difficulties students faced with that material, and I have spent some time thinking about good ways to explain it.

That course reinforced Bob's desire to pursue mathematics more intensively, and he did indeed graduate with a master's degree. I spent some time with Bob and realized that he had the ability to absorb mathematics in whatever form it was presented as well as come up with insights on his own. Bob was somewhat older than the typical graduate student, and as a result, pursuing a doctorate was probably not a reasonable option for him, although I have no doubt he could have obtained one.

Bob's background was not a standard academic one. After obtaining his degree, he and his family started several businesses. The first had not done so well, but the second was thriving. It consisted of a chain of several Hooters-type restaurants and bars. The food and drink were standard sports-bar fare, but the primary attraction, like Hooters, was the waitresses.

The waitresses were young women, and every so often, they would be unable to obtain transportation to and from their place of work. Bob's family took care of this by transporting waitresses who ran into such difficulties, and Bob asked me if I would like to accompany him one evening when he was doing this. Although offering such an invitation may seem unusual for a student to offer a professor, Bob wasn't taking classes from me and occupied a somewhat nebulous middle ground between student and acquaintance.

Bob spent most of the time that he wasn't in the car by the telephone. Every so often, I'd hear his end of the conversation, and we'd go pick up Carol to go to Bar #3 or Susie to go to Bar #1. But once in a while, he'd receive several calls at once and had to do a bit of planning as to what order to pick up the girls and drive them to the various bars in order to do this efficiently. All in all, although I didn't obtain any telephone numbers of single young women interested in dating a math professor, it was an enjoyable evening.

When Bob graduated, I wrote a recommendation letter for him and was happy to hear that he had obtained jobs teaching at community colleges and universities. Bob and I kept in touch infrequently and informally over the years.

I retired from my teaching position at CSULB in 2013. At about that time, I learned that Bob had encountered some problems, resulting in campus police authorizing a 5150 hold to be placed on him. Section 5150 of the California Welfare and Institutions Code authorizes a qualified officer or clinician to involuntarily confine a person suspected to have a mental disorder for 72 hours. I also learned that Bob had been dismissed from a teaching position at the institution which had authorized the 5150 hold. Bob and I had never been close, and although I was sorry to hear about it, I did not contact Bob – nor did he contact me then. But he did contact me a few months later.

The Traveling Salesman Problem

During the spring of 2014, I received an email from Steve (also not his real name), who identified himself as a friend of Bob's. He said that he and Bob had become interested in the traveling salesman problem (TSP), and they believed they had achieved a breakthrough, which they would like to discuss with me over lunch at a restaurant close to CSULB. Despite the difficulties Bob had apparently encountered a few months earlier, I had always enjoyed his company, and I knew he was capable of generating interesting ideas, so I had no problem accepting the invitation. I told Steve that although I knew what the TSP was and a little something about it, I was by no means an expert. He said that even though I was not an expert, I would almost certainly find what they had to say extremely interesting.

The TSP is easy to describe. A salesman residing in a city has a number of other cities to visit, after which he returns to his hometown. The goal is to minimize the total distance traveled. This problem is often discussed in elementary courses on combinatorics, where it is easy to show that if the salesman has N different cities to visit and is required to start from a specific one of those cities, there are $N!$ different possible routes that he might take. $N!$ is a shorthand for the product $1 \times 2 \times 3 \times \cdots \times N$, and these numbers become large very quickly. $5! = 120$, but $10! = 3,628,800$, and $20! = 2,432,902,008,176,640,000$.

I can summarize what I knew about the TSP prior to the meeting with Bob and Steve. I knew that finding the optimal route was a problem of exceptional difficulty and that its solution would mark a major development in mathematics. At the turn of the twenty-first century, the Clay Institute offered $1,000,000 for solutions to seven of the unsolved problems in mathematics, and the TSP was intimately involved in one of those problems.

But beyond the $1,000,000 offered by the Clay Institute, I also knew that the solution to the TSP would have significant financial value. Finding the best way to do something can obviously be very important, and there were a number of major problems related to the TSP. You encounter one of these every time you plan a vacation. Suppose you have decided to visit six cities. If you are on a budget, you have a lot of work to do – balancing the cost of airplane travel and the hotel rates in the different cities. As you know, the cost of an airplane trip can vary substantially depending on when the plane leaves, and the cost of staying at a hotel can be very different on a weekend from what it is on a weekday. Now, imagine that you are a company whose employees have to do a lot of traveling in order to conduct business. You can see how knowing how to do this in the most economical fashion could save the company a lot of money.

Many great developments in math and science arise from situations that are encountered in the real world. Most people have heard about how an apple falling on Isaac Newton's head kicked off what would eventually become Newton's Theory of Universal Gravitation. This story may be apocryphal, but one that doesn't involves how Richard Feynman's discoveries concerning relativistic electrodynamics were stimulated by a problem he encountered in the cafeteria at Cornell University. Someone threw a spinning plate into the air, and Feynman decided to solve how the pattern engraved on the plate behaved as the plate spun through the air. After he had solved this problem, he realized that it had connections to the way an electron's orbits evolve in the theory of relativity. With these in mind, I couldn't help but think that perhaps Bob had been motivated to look at the TSP because he had a previous exposure to it when he tried to work out the most efficient way to take the waitresses to their places of work.

I knew that there were some subtleties involved in solving the TSP. There is an obvious algorithm: Start by going to the nearest city, and then just go to the closest city yet unvisited. This is known as the nearest neighbor algorithm, which can be shown to be defective using the following simple

Table 9.1.

	Home	A	B	C
Home		100	101	102
A	100		50	200
B	101	50		300
C	102	200	300	

example. The entries in Table 9.1 are simply the distances from one location to another. For example, the distance from *B* to *C* is 300.

Here's what we could obtain by using the nearest neighbor algorithm, If we start from Home, the place nearest Home is *A*, so we go there. From *A*, the nearest unvisited location is *B*, so we go there next. This forces us to go to *C* because we haven't been there yet, and from there, we go Home. The total distance is $100 + 50 + 300 + 102 = 552$.

But there's a much shorter route. Start from Home, and go to *C* first, then to *A*, then to *B*, and then back Home. The total distance is $102 + 200 + 50 + 101 = 453$. The nearest neighbor algorithm is what is known as a greedy algorithm because it only looks one step ahead. One can improve the nearest neighbor algorithm somewhat, but at a cost – you have to do more computation.

And that's why the TSP is so difficult. The number of possible routes grows factorially. 20! is about 2 quintillion, but we're into the sextillions with 23!.

One of the reasons that the TSP is so important is that its solution will unlock solutions to a number of other problems of immense practical importance. One of these problems is known as the knapsack problem, which involves the most efficient way to pack things. You can see how something like this could be tremendously useful.

When I met with Bob and Steve, they outlined their proposed solution, which they described in terms of car trips. Cars would start out in the salesman's hometown, and a fleet of cars would head for each of the cities, all traveling at the same speed. When a car reached a city, it would prompt a fleet of cars to go to all the unvisited cities, and this process would continue, with all the cars traveling at the same speed. The first car to return home would obviously have traveled the shortest distance.

I went home and thought about this, but soon realized that this was simply examining all possible trips, replacing distance with time using the formula DISTANCE = RATE × TIME, which everyone learns in school. Of course, it would have been prohibitively expensive, as a really huge fleet of cars would have been required, but one could simply send messages rather than cars to reduce the expense. I relayed this information to Steve and Bob, but encouraged them to contact me again if they had another idea to run by me.

I did not hear from them again until April 2017. Again, I received an email from Steve, but this time, he said that he and Bob had an extremely exciting development that they were sure would interest me. We met again for lunch, and they had obviously become a lot more familiar with the problem. They told me that the first major step toward the solution of the TSP had been taken by the mathematician George Dantzig in 1954. Dantzig had decided to tackle the problem of finding the shortest route which visited all the state capitols, and he had devised an algorithm which had, indeed, found the shortest route which visited 42 state capitols. Dantzig had restricted the problem to visiting the capitols of the 48 states in the continental United States and, for convenience, had decided to treat several of the capitols of the northeastern states, which were close to one another, as a single city.

Over the years since 1954, many accomplished mathematicians and computer scientists had tackled the problem, and there was extensive literature on the subject. A number of extremely sophisticated algorithms had been devised, and with the advent of high-speed computers, there were some remarkable developments. The optimal routes were known for a number of examples, and new algorithms and new computational techniques were often tested by seeing how they performed on these known collections. One of these examples was the Dantzig 42, and every serious algorithm developed to date had come up with precisely the same solution as Dantzig did in 1954.

Of course, none of these algorithms had produced its route by examining every one of the 41! different routes available, as 41! is on the order of 3×10^{49}, well beyond the capability of any computer to evaluate directly.

So, even though Dantzig had proved mathematically that his route was the shortest, proofs are sometimes shown to be erroneous. And even though the subsequent algorithms that worked on the Dantzig 42 had come up with the same result, without an expert on algorithms there was no way to know whether the subsequent algorithms incorporated the same basic technique as did Dantzig's original algorithm. I certainly had no idea if this was the case.

What Steve and Bob claimed to have done was, to me, astounding. They believed they had developed an algorithm which came up with a shorter route through the Dantzig 42 than the one that Dantzig found! There was absolutely no question that, if Steve and Bob's result stood up, it would have been a bombshell. The TSP was one of the most famous unsolved problems in mathematics. Discovering a flaw in a work that had been accepted for 50 years, especially in a problem of this magnitude, would have been of monumental importance.

To substantiate their claim, they handed me a printout from an Excel spreadsheet that compared their route with the Dantzig 42. At first glance, it looked correct to me – but of course, I wanted to check their calculations for myself. Looking at that sheet, I could understand their excitement. If correct, it would not only be a bombshell in and of itself, but the algorithm that had unearthed it could also conceivably be worth a fortune.

When I went home to check, the first thing that occurred to me was that, possibly, the difference between their result and the Dantzig result might have been due to accumulated rounding errors. On looking at their sheet, I was quickly able to eliminate this as a possibility – the difference between their result and Dantzig's was too great to be accounted for by rounding errors. I checked their calculations more thoroughly, and it seemed to me that their calculations were indeed correct. The route they had found was significantly shorter than the one Dantzig had found, and every subsequent well-regarded algorithm had confirmed.

I'm not sure that it's possible any more for a significant breakthrough to be made in the natural sciences by a relative amateur. Yes, every so often, one reads about an amateur astronomer being the first to spot an exciting development, but this is mostly a matter of being fortunate enough to look in the right direction at the right time. To the best of my knowledge, all of the major scientific breakthroughs nowadays are made by highly trained scientists, and often the breakthroughs require several teams working on different aspects of the problem.

But mathematics is different. As we saw in the previous chapter, Yitan Zhang, a mathematician who at the time was managing a Subway franchise in order to make ends meet, made a major discovery on the twin primes conjecture. Zhang was virtually unknown to the mathematics community, and his ideas were completely novel, as well as the result of almost a decade of thought. This had occurred just a year before Steve asked me to have lunch with him and Bob. And it was with that example in mind that I continued to check on what could be a development even more astounding than Zhang's. Extraordinary claims require extraordinary proof, but the proof here wasn't really extraordinary; it was simply a matter of checking computations.

I rechecked their computations several times, doing it both with Excel and by hand. And then I noticed something that struck me as a little odd. Steve and Bob's computations had distances between the cities correct to the nearest hundredth of a mile because they were using a distance formula from geometry that used the map coordinates of the individual cities.

My first real job was programming computers during the summer while I was in college in the early 1960s. FORTRAN, the computer language that I used at that time, distinguished between integers and numbers, which were expressed using decimals, and it was felt desirable to express as many quantities as possible as integers because the computer could do integer calculations significantly faster than ones involving decimals. So, it occurred to me that possibly Dantzig had rounded distances to the nearest mile because it would have made for quicker calculations. I decided to check and went back to Dantzig's original paper.

Dantzig had indeed stated the distances in integers. Reading his paper more carefully, I discovered that Dantzig had obtained his distances between cities by using a Rand McNally road atlas from 1954. Red alert! This raised the serious possibility that the distances that Dantzig had used were significantly different from the distances that Steve and Bob had used. After all, the geometric formula that Steve and Bob used gave the actual physical distance between two cities as measured on what is usually referred to as a "great circle route". A geometric plane is determined by three points, and if the three points are the two cities and the center of the Earth, the plane containing these three points will intersect the surface of a spherical Earth in a circle – referred to as a "great circle". Imagine the face of a circular clock, and you want to travel from 12 o'clock to 4 o'clock along the circumference of the circle. There are two routes, which are arcs of a circle: The short one goes via 1, 2, and 3 o'clock to 4 o'clock, and the long one goes backward, through 11, 10, and 9 o'clock. The "great circle route" between two cities is simply the shorter of the two circular arcs; it is the shortest of all possible paths between two cities on the surface of the Earth.

However, as is well known, roads between two cities do not always take the shortest path. This difference is critical to several well-known algorithms, which, like the nearest neighbor algorithm, can be highly sensitive to small differences. Although there was no way I could tell for sure, the obvious thing to do was to check how Steve and Bob's algorithm fared using the distances that Dantzig had used in his original paper.

I suggested as much to Steve and Bob. If indeed their algorithm came up with a shorter path than Dantzig's using Dantzig's original data, it would have been a truly sensational discovery. I also mentioned to Steve and Bob that even if their algorithm did not give a shorter path but gave the same path as the one Dantzig had discovered, it would mean that their algorithm could very possibly be of value. I was also careful to state that this was not my area of expertise but that there were a lot of experts on the TSP, and they should perhaps talk to some of them.

I did not hear from either Steve or Bob for almost a year. All of my contacts with them concerning the TSP had been made through Steve, as Bob was not very computer-savvy at the time, and his inbox was either neglected or full. I had occasion to contact Steve a year later, when I was giving a talk at CSULB, which I thought might be of interest to them.

During the course of our exchange, Steve told me that he and Bob initially had a difference of opinion on what to do with the algorithm, and the difference of opinion had expanded into a full-scale rift. Bob had inundated Steve with emails and phone calls, some of which Steve interpreted as threatening, and finally Steve had to file a restraining order, which had been granted. Of course, I was sorry to hear it; Steve had impressed me as a straightforward guy, and Bob and I had a relationship extending back more than 30 years.

It was fairly obvious that the falling-out probably involved the failure of the algorithm in one way or another, and I didn't think much more about it. A few months later, though, I received an email from Bob. He said that he was involved in a legal dispute with Steve, and he wanted me to give a statement regarding my experience concerning their algorithm to the judge. I crafted one that satisfied both of us and felt the matter was at an end.

The hearing with the judge did not go Bob's way, and he asked me for copies of every email I had exchanged with Steve. Because some of them were inconsequential, I had deleted them, although I had kept the ones relevant to my discoveries and explorations concerning the Dantzig 42. However, it wasn't clear to me that Bob was entitled to them, and as he and Steve were involved in some sort of litigation, I felt I might be incurring some sort of legal jeopardy by turning over the emails. I told Bob this but assured him that there was absolutely nothing in those emails of which he was unaware, and I had no problem if he obtained them from Steve.

As events unfolded, it became clear that Bob had come to the conclusion that Steve had interpreted my remarks as implying that I considered the algorithm to be valueless. I had never even come close to saying anything like that and had specifically pointed out that, even though I had no idea what the algorithm actually was and even if it did not reveal anything new about the Dantzig 42, it could still be valuable in one of several ways. It could be faster than existing algorithms, or it could handle problems others couldn't. There was no way I could tell, as I have no expertise in algorithms and had, at any rate, never seen their algorithm. Steve and Bob were joint owners of the algorithm, and an impasse had developed because Bob wanted to do something with it that Steve was reluctant to do.

What I had failed to reckon with was that Bob would take my refusal to hand over the emails as proof that Steve and I had come to some sort of arrangement with regard to the algorithm. I assured Bob that this was not the case, but despite our long association, he did not believe me. I would receive emails and phone calls from Bob, just as Steve had. Some were interesting, dealing only with mathematics or physics, and I could tell that Bob was still capable of coming up with intriguing ideas. However, the threats became more intense, and when the threatening behavior extended to involving my wife, I could no longer assume that rational discourse would suffice and that I could talk Bob off the ledge. I looked up the requirements for filing a restraining order on the grounds of civil harassment and did so.

On a gloomy day in December 2018, Bob – accompanied by his lawyer – and I met in a courtroom. I had submitted a two-page statement, along with copies of some of the more obscene and vitriolic emails I and my wife had received. We were instructed to exchange documents, and I saw that Bob's lawyer had prepared an eight-page summary of the information he received from Bob. That document contained many incorrect statements that I was prepared to challenge if need be. His lawyer and I met outside the courtroom before the time came for the judge to render a verdict. The lawyer asked me if I would be willing to withdraw the complaint, and I said that I could not, as I had no guarantee that the harassing behavior, which had greatly upset my wife, would stop.

When we were called back into the courtroom, the judge – a no-nonsense woman whom I judged to be in her early 50s – asked Bob if he had written a particularly vitriolic email to my wife, which I had included in my statement. He couldn't very well deny it because it clearly came from his email address. The judge told Bob that he couldn't write threatening emails like that and granted all provisions I had requested in the restraining order.

In the interim, I had had several conversations with Steve, who had washed his hands of the whole affair and who had said he had no problem with my sending the emails we had exchanged to Bob. In retrospect, it would doubtless have been easier had I done so earlier, so certainly some blame rests with me. However, had Bob been reasonable, he would have accepted my assurance that there was nothing in the emails that Steve and I had exchanged that would cause him any problems. At any rate, I sent Bob's lawyer the emails, as I did not want to communicate with Bob directly. Bob obeyed the injunctions of the restraining order, and my wife has not been harassed since the hearing, which was my primary concern.

But how could this have happened? It was clear – at least to me and Steve – that I had done both of them a huge favor by finding the potential flaw in their presentation. Bob had wanted to announce this result in some publicity-garnering fashion and believed that the algorithm was worth millions of dollars. Certainly, if it had shown an error in the Dantzig 42, it would have brought the two of them fame and, very likely, fortune. However, I had prevented them from making fools of themselves because they had overlooked something totally obvious. In the academic world, you can recover from an error – we all make them – but it is impossible to recover from a publicity-seeking error. As evidence of this, consider the brouhaha surrounding the announcement in 1990 of cold fusion by the chemists Martin Fleischmann and Stanley Pons. Had Fleischmann and Pons just written a paper with their presumed results, it would have undergone peer review, and people would have tried to duplicate their results and failed (which actually happened after they announced their presumed results). Their careers and reputations would have suffered no damage. We all make mistakes, and science has a way of finding out about those mistakes through experiments that attempt to duplicate the results. But you don't want to make mistakes in such a way as to bring potential discredit to yourself and your institution on a national stage. I had prevented Steve and Bob from committing the same type of professional suicide that Fleischmann and Pons had brought upon themselves. At least, that was how I saw it.

But Bob, apparently, saw it differently – truly it is written that no good deed goes unpunished. Because Steve was no longer interested in working on the algorithm as a result, I suppose, of what he had discovered when he tried applying the algorithm to the Dantzig 42, Bob may have felt that I was responsible for breaking up the partnership. In addition, he also seemed to have a paranoid suspicion that Steve and I had some sort of arrangement to deprive Bob of the fame and fortune to which he felt he was entitled. In vain, I tried to point out to Bob that since he and I had a professional and previously amicable relationship extending back decades, I would be much more likely to form an arrangement with him than with Steve, whom I knew only through a couple of lunches and email exchanges. I felt that whatever psychological problems had caused authorities to originally issue the 5150 hold had intensified in the five or six years that had elapsed.

Bob was – and possibly still is – a very bright and talented scientist. But other scientists, even more talented, had wrestled with demons much more terrifying than Bob's. In the course of writing an earlier book, I had come across two of these: Ludwig Boltzmann and Wallace Carothers. And, of course, many people know about John Forbes Nash, whose story Sylvia Nasar told so brilliantly in *A Beautiful Mind*.

Other than math coming more naturally to me than to most people, I think I'm pretty normal. I get unhappy occasionally – who doesn't – but it never comes remotely close to depression. When I'm unhappy, I don't work very effectively; it affects my day-to-day behavior and my performance of any number of tasks. This is especially true when I have tried to do serious mathematics; my mind simply will not focus on mathematics and gets distracted by the personal problems I am experiencing.

But people such as Boltzmann, Carothers, and, of course, John Forbes Nash must have been tremendously courageous and focused people to be able to overcome, if only for limited periods of time, the psychological conditions which would totally hamstring many of us. I've seen depression in both my family and friends, and it's hard for me to imagine how a scientist who is depressed can do great science. I know that many artists, musicians, and writers have been extraordinarily creative, though beset by psychological problems, but creativity in these areas seems very different to me than creativity in the sciences, Possibly Hieronymus Bosch was mad, and this madness led him to paint *The Garden of Earthly Delights* – a work that strays considerably from reality. But to be a top-notch scientist, you have to focus. There's an objective reality out there, and you must come to grips with that while battling your demons.

Bernhard Riemann's Housekeeper

As I mentioned, at the hearing with Bob, the judge granted my requests but then asked me for how long I wanted them to remain in force. Although I certainly didn't want my wife to be harassed, all I wanted to do was get on with my life, and I believed Bob wanted to get on with his. So, I asked for six months.

As I mentioned, Bob obeyed the injunctions of the restraining order, but a few weeks after the six-month period expired, I received a reasonably cordial email from Bob discussing some interesting questions in math and physics. I answered, and we began a correspondence, but after a few back-and-forths,

Bob again started making accusations and threats regarding the experience with the algorithm. I now realized that this behavior was probably not going to change, and so I told Bob that I was not going to read or respond to his emails, but I would keep them in case I needed to file another restraining order.

However, when someone I know sends me an email, I'm tempted to read it, and after that, I'm almost always tempted to respond. In order to avoid going down that particular rabbit hole again, I created a hidden folder which is not visible to me when I initially look at my emails. My email server allows me to direct emails from a particular sender to a folder, so I directed Bob's emails to that hidden folder.

But every so often, I open Pandora's box by looking at the folder. I just looked at it before writing this; it now contains 602 unread emails, the last batch being sent in August 2021. Although I did not read any of the emails, I did read the subject lines. Some were threatening; there were a few with the subject line "Why you will be sued", but I haven't been yet and doubt that I will be. Bob is probably more than 80 years old, seriously overweight, and has always been much stronger on thoughts than on actions.

But some of the subject lines in those emails tempted me to read them. One subject line reads, "Infinitesimal rotations are NOT commutative, contrary to most physics books I have read". The temptation to dive in is strong, but I know where that will end up, and I'm not going there.

Bernhard Riemann was one of the great mathematicians of the nineteenth century. Like most mathematicians, he enjoyed the process of doing mathematics a lot more than the process of writing up his results for publication. When he died, his housekeeper disposed of his unpublished notes and, in doing so, possibly deprived the world of results of considerable importance. With this in mind, I am going to preserve Bob's unread emails – and any future ones that may show up in that hidden folder – because infinitesimal rotations may *not* be commutative, and I don't want my legacy to be the same as that of Bernhard Riemann's housekeeper.

Part IV

Love Affairs

When my first book was published, I was contacted by Marshall Poe, the founder of New Books Network, an internet website devoted to – as you might guess – new books. He asked me if I would be willing to conduct podcasts on new books in math and science for the New Books Network. I jumped at the chance, as it would enable me to talk about two of my favorite topics – math and science – with people whom I never would have met otherwise.

One of those people was Edward Frenkel, the author of *Love and Mathematics*. His book talked about his personal life and how it intertwined with the problem that was the quest on which he had focused. That problem is known as the Langlands Program, and although it wasn't an area of interest to me, I could certainly relate to Frenkel's passion for it, and I enjoyed discussing his book and his passion with him.

So, the last part of this book is about three of my love affairs with mathematics. The first is the Banach Contraction Principle, which occupied the second half of my mathematical research. The second is about the multi-armed bandit problem, on which I'm currently working.

And the third and last is on poetry – and yes, mathematics enters into it, and no, it's not poems about mathematics.

Chapter 10

The Banach Contraction Principle

After my father passed away, I needed to focus on reviving my career as a university professor. It hadn't been an easy decision to relinquish my tenured position, but it had been necessary to do so in order to cover my father's medical expenses. I hoped to return to the university environment, as I liked both teaching and research, and I liked having a lot of time to myself. Where else other than a university do you have this available?

Fortunately, I was still considered to be a relatively valuable addition to the mathematics department at CSULB. I think in this aspect – and in many others – I've been tremendously lucky. I went to Yale as an undergraduate; they now accept 1 of every 20 applicants, and since I've served on alumni interview panels, I'm pretty sure that I wouldn't make the cut. Likewise with going to grad school at the University of California at Berkeley and getting a tenured position at a university in the Los Angeles area. A couple of years ago, CSULB had tenure-track positions available and received on the order of 400 applications from all over the world. My application might not have survived the initial part of the weeding-out process.

But it was 1988 rather than 2024, and I had an "in" at CSULB. I was first rehired as a part-timer, then on a one-year contract as a full-time instructor, and finally on a tenure-track position. I also discovered that the five or six years that I had been out of the research game had made it virtually impossible – as well as unappetizing – to continue research along the lines that I had originally started when I first obtained my doctorate.

I mentioned in the chapter on the incompleteness theorem that the continuum hypothesis changed my life after I came back from my options trading interlude. Not only had I lost valuable time doing research – and five years is a lifetime when it comes to cutting-edge research – but I found it philosophically unappealing to work in an area in which the problems depended on whether or not you were willing to incorporate the continuum hypothesis in your axiom set. I badly needed a different problem on which to work. I found one thanks to a Polish mathematician whose research impacted mine at the two critical stages of my academic career: before I left to trade stock options and after I returned.

Stefan Banach

Werner Heisenberg, after whom the uncertainty principle in quantum mechanics is named, once said that the world of the atom is not only stranger than we imagine, but it is stranger than anything we can imagine. And if there is one theorem in mathematics that falls under the heading of "stranger than we can imagine", it would be the Banach–Tarski Theorem.

The Banach–Tarski Theorem states that it is possible to take a sphere the size of a pea, disassemble it into a finite number of parts, move these parts around, and then reassemble them into a sphere the size of the Sun! Now, before you run out to buy a small ball of gold and try to reassemble it into a much larger ball of gold, you need to understand that the disassembly and reassembly processes cannot be accomplished in the real world, and the finite number of parts referred to above don't look like chunks but more like clouds or swarms of particles. And also, even if you could by some chance do this (and you can't), gold is sold by weight, not volume, and the weight of the large sphere would be the same as the weight of the small sphere. Nonetheless, it is an absolutely startling theorem and worthy of an entire book, and a very good one, *The Pea and the Sun*, has been written on it by my friend Len Wapner.

But most mathematicians, and certainly those in the areas in which I worked, would say that the Banach–Tarski Theorem is sort of an *amuse bouche* when compared with the meat and potatoes of Stefan Banach's mathematical work.

Two fundamental areas of study in mathematics are algebra and topology. Most people are exposed to algebra fairly early in their progress through school, learning how to add, subtract, multiply, and divide polynomials. And that's one of the basic things that algebraists do; they study systems with these familiar arithmetic operations. Topology is a little more difficult to describe, but for the systems in which Banach specialized, the important thing was that one could define the distance between two objects in the system. Banach realized that there were a large number of important mathematical systems which incorporated both algebra and topology, and his name is attached to some of the most important of these systems, such as Banach spaces and Banach algebras.

Banach algebras were of great interest to Bill Bade, my thesis advisor, and the problem on which he assigned me to work toward my doctorate – the uniqueness of norm problem – involved Banach algebras. So, there's the obvious connection between Banach and me from the moment I started on a career in mathematics. And in the first half of my career, one of my (few) high points was coming up with a beautiful theorem that generalized the uniform boundedness theorem, one of the three most important theorems in functional analysis. I say "beautiful" because it really is – and much of the beauty is due to Vlastimil Pták, a Czech mathematician whom I never met but who came up with the basic idea and method of proof that I used to obtain the result. And the reason this is being mentioned here and now is because the uniform boundedness theorem is also known as the Banach–Steinhaus Theorem. So, it's fair to say that most of the first half of my mathematical career consisted of going down paths that Banach was one of the first to explore.

The Banach Contraction Principle

The Banach–Tarski Theorem has nothing to do with Banach spaces or algebras; it's just another example of the scope and brilliance of Banach's mathematical accomplishments. And the same can be said for the Banach Contraction Principle, which is taught in introductory courses in real analysis, usually encountered by math majors after they have completed a three-semester calculus sequence. There's a good reason for this. The development of calculus, in terms of techniques, took place much faster than the theoretical understanding of exactly when and why these techniques were successful. As a result, the situations in which they were unsuccessful weren't really studied and analyzed until it became apparent that there were such situations and that they were very important.

You can skip most of this paragraph if you never took a calculus class, but I want to include it because it's worth seeing for those who have. It's fairly easy to understand the Banach Contraction Principle as it applies to real-valued functions of a single real variable – the stuff one comes across in first-semester calculus. Suppose that f is a real-valued function of a single real variable x and that there is a constant M with $0 < M < 1$ such that $|f(y) - f(x)| \le M|y - x|$. Functions satisfying this property are called contractions because the distance between the two points x and y contracts by a factor of at most M when f acts on these two points. The inequality above can be stated in English as the distance between $f(x)$ and $f(y)$ decreases by a factor that is at most M from what the distance between x and y was.

Any function with derivative $|f'(x)| \le M$ (where $0 < M < 1$) is a contraction. It's a consequence of the mean value theorem, a result which generally goes unappreciated by students just beginning a course in calculus. It says that if f is differentiable on the interval (a, b), there is a point c in (a, b) such that $f'(c) = (f(b) - f(a))/(b - a)$. The mean value theorem has an interesting interpretation when considering velocity. It states that if your average velocity on a trip was 40 miles per hour (if $f(t)$ is your position at time t, $(f(b) - f(a))/(b - a)$ is your average velocity between times a and b), then at some point during the trip, your speedometer would have said 40 miles per hour (instantaneous velocity is the derivative of position, so $f'(c)$ is your instantaneous velocity at time c).

Getting back to the Banach Contraction Principle, if f is a function whose derivative satisfies $|f'(x) \le M$, where $0 < M < 1$, then on the interval (a, b), there is some point c between a and b for which $f'(c) = (f(b) - f(a))/(b - a)$. So, $|f(b) - f(a)| = |f'(c)|b - a| \le M|b - a| < |b - a|$, and hence f is a contraction.

What the Banach Contraction Principle says is that every contraction has a unique fixed point. A fixed point for a function f is a number x such that $f(x) = x$. Geometrically, a fixed point is where the graph of the function crosses the line $y = x$. Some functions don't have fixed points;

$f(x) = x + 1$ is an example, because if it had a fixed point x for which $f(x) = x$, then $x = f(x) = x + 1$.

My favorite example of a fixed point occurs when one converts Fahrenheit to centigrade (or Celsius, if you prefer) by means of the formulas $C = 5(F - 32)/9$. Note that the Banach Contraction Principle states that the function $f(x) = 5(x - 32)/9$ has a unique fixed point (because its derivative is 5/9), but it doesn't tell you what that fixed point is. It's easy to find by solving the equation $5(F - 32)/9 = F$; the solution is $F = -40$. $-40°$ is the same temperature whether it's $-40°F$ or $-40°C$.

It's easy to show that if a contraction has a fixed point, that fixed point is unique, i.e. there's only one fixed point. Suppose f is a contraction with a constant M (where $0 < M < 1$), and assume that a and b are two distinct fixed points. Then, $|a - b| = |f(a) - f(b)| \leq M|a - b|$. If a and b are distinct, then $|a - b| > 0$, so we can divide the previous inequality throughout by $|a - b|$ to get $1 \leq M$, an obvious contradiction.

That's the easy part of the proof of the Banach Contraction Principle. It was easy to find a solution to $5(F - 32)/9 = F$; that's just simple algebra. But the function $f(x) = \cos(3x/4)$ is also a contraction, and how do you solve the equation $\cos(3x/4) = x$?

As you might suspect, you can't solve it – at least, you can't solve it exactly. But Banach was able to show that a solution exists, and he gave a really cute procedure to solve it – at least in theory.

Suppose we have a contraction $f(x)$. Pick a starting point x_0 (any starting point will do). Now, compute $x_1 = f(x_0)$, $x_2 = f(x_1)$, $x_3 = f(x_2)$, etc. Knowing x_n, we compute $x_{n+1} = f(x_n)$. This is known as a recursion formula. What Banach was able to show was that the sequence $x_1, x_2, x_3, \ldots$ gets arbitrarily close to a particular number (that's the topology part), and that number is the fixed point. Let's see how this works for $\cos(3x/4)$. Because my wife's birthday is September 1st, I'll use 9.01 as x_0. Table 10.1 shows the recursion computation up to x_{13}.

Make sure your calculator is set to radians, and compute cos(3/4 (0.81775)). The answer, to five decimal places, is 0.81775.

Fixed points are important in many physical processes and elsewhere – just ask John Forbes Nash, who received a Nobel Prize in Economics for his work on fixed points, as well as a book (*A Beautiful Mind*, by Sylvia Nasar – highly recommended) and a movie with the same title.

Table 10.1.

n	x_n	n	x_n
0	9.01	7	0.818224
1	0.889606	8	0.817541
2	0.785555	9	0.817836
3	0.831404	10	0.817709
4	0.811809	11	0.817764
5	0.820303	12	0.81774
6	0.816643	13	0.81775

The Only Original Idea I Ever Had

So here I was, looking for a quest on which to embark. When you're addicted to math, you want to do math. In a recent *Science* magazine, a number of scientists were asked why they liked doing research. One gave exactly the answer that I would have given: When you discover a result, there's this one transcendent moment that you think, "No one else in the Universe knows this but me". The rush I experience in these moments far exceeds any rush I have ever experienced in any of the many games I have played. I remember that I always wanted to win a national tournament in bridge; I did, even though it was a relatively minor event, sort of like winning the Over 45 Mixed Doubles at the U.S. Open in tennis. But, hey, it's a national championship! At any rate, I had a brief moment of celebration with my teammates and woke up the next morning thinking not "I'm a national champion", but "Well, what's next?"

Then, while giving a lecture on the Banach Contraction Principle in an upper-division analysis class, I had an idea that revived my research career. In the real world, sometimes one person does a job, such as painting a house. Sometimes the job is shared among a number of people, each being responsible for painting several rooms. But in mathematics, I had never seen a situation in which the "job", which is generally a hypothesis that a mathematical object must satisfy, is shared among several objects.

It doesn't require much to understand the idea that occurred to me. I'm going to replace the function notation $f(x)$ with the operator notation Tx. It's still the same thing, but this enables a succinct way to compose a function with itself. $T^2 x = T(Tx)$, and recursively, $T^{n+1}x = T(T^n x)$. I'm using the operator notation here so as not to confuse superscripts with differentiation, which might happen if I used function notation.

The key hypothesis of the Banach Contraction Principle is that $|Tx - Ty| \leq M|x - y|$. Here, the entire "load" of the hypothesis rests on the shoulders of T. What if T and T^2 share the work? By this, I mean that for every pair (x, y), either $|Tx - Ty| \leq M|x-y|$ or $|T^2 x - T^2 y| \leq M|x-y|$?

This was ostensibly a weaker hypothesis than the one in the Banach Contraction Principle, but first I had to show that there really were functions T which satisfied the weaker hypothesis but not the stronger. Maybe I didn't really need to do this, but I remembered that one of my professors at Yale had spent two years investigating the properties of a particular mathematical system – only to have it shown later that there were no such systems! Sort of like a biologist spending two years studying what properties one could expect of unicorns, only to have it demonstrated that there are no such things as unicorns. I certainly didn't want to find myself in that position.

Once that was done, the next order of business was to research the literature to see if anyone had worked on this problem. Apparently, no one had, so the field was completely open. It's a lot easier to obtain publishable results when you're the only one looking at a particular problem.

Earlier, I talked about the rush you experience when you discover something that possibly no other entity in the Universe knows. I'd experienced that previously during the first half of my academic career, but any such rushes were slightly tainted by my knowledge that there were better mathematicians than I working on the problem on which I was working, and even as I was writing my results up for publication, the next edition of some journal might show that others had gotten there before me. But in this case, I could be pretty sure that no one else had gotten there before me because no one else had even thought of investigating this problem.

I was filled with enthusiasm at the thought of working on a problem no one had even considered before. It occurred to me that my original idea of having T and T^2 share the load of the contraction hypothesis had an obvious generalization: Given an integer N, have $T, T^2, T^3, \ldots, T^N$ share the load. That would be a really lovely theorem if I could prove it. But one of the things that I, and many other mathematicians, have learned through grim experience is not to bite off more than you can chew. Always see if you can prove an easier case first.

The case of $N = 2$, where the load is shared by T and T^2, turned out to be quite interesting. I needed to construct some combinatorial arguments, and with the help of Bruce Rothschild, a long-time friend from my UCLA days, we managed to do it. Having Bruce as a co-author had an unexpected side benefit: I acquired an Erdös number of 2.

Erdös Numbers

I am proud to say that I am one of about 17,000 mathematicians to achieve an Erdös number of 2. Paul Erdös was the most prolific mathematician of the twentieth century and one of the best; he's the only mathematician I know of whose death was the subject of front-page articles in major newspapers. Erdös would travel from university to university, staying with colleagues for a couple of weeks, giving talks, and doing research. I like doing research, but there's not much evidence that Erdös liked anything else. At any rate, someone devised the idea of an Erdös number, indicating the closeness of the author linkage with Erdös. Erdös himself had an Erdös number of 0, any mathematician who co-authored a paper with Erdös had an Erdös number of 1, and Erdös numbers are recursively defined by the following statement: A mathematician receives an Erdös number of N if he has co-authored a paper with a mathematician having an Erdös number of $N - 1$ and has not co-authored a paper with a mathematician having an Erdös number of $N-2$ or lower. Bruce Rothschild co-authored a paper with Erdös, one of about 500 people to do so, and thus has an Erdös number of 1. Having co-authored a paper with Bruce and not having co-authored a paper with Erdös, I have an Erdös number of 2.

Of course, the idea of an Erdös number goes back to the "six degrees of separation" concept: the idea that any two people in the world can be joined by a chain with at most six links. In this case, a link is a connection representing the fact that the two people know each other. The chain Alice–Bob–Charles–Donna–Evelyn has four links. Alice knows Bob, who knows Charles, who knows Donna, who knows Evelyn. Assuming that this is the shortest chain that can be formed from Alice to Evelyn, there are four degrees of separation between Alice and Evelyn.

A Surprising Reaction

I had an upcoming sabbatical, and one of the things I wanted to do was go up to Berkeley and talk about this with Bill Bade, my thesis adviser. I was quite pleased with the way things were going: I had published a paper on it in a major journal, I had obtained some interesting further results, and I had realized the idea of "sharing the load" had potential application to almost every area of mathematics.

Here's an example. A function T is said to be additive if $T(x + y) = T(x) + T(y)$ for all x and y belonging to X, where X is a set of some sort, perhaps a vector space. The "job" is specified by the hypothesis that $T(x + y) = T(x) + T(y)$ for all x and y belonging to X, and the only "worker" assigned to the job is T. But what if we had a collection of functions $\{T_a : a \varepsilon A\}$ such that for every pair x and y belonging to X, we could find some T_a belonging to the collection such that $T_a(x + y) = T_a(x) + T_a(y)$? The "job" would be shared by the collection of "workers" $\{T_a : a \varepsilon A\}$.

When I discussed this idea with Bill, his reaction was totally unexpected. In the thirty-plus years that I had known him, I had never seen him upset. After I had explained the idea and the extension I had proved for the well-known theorem, he said with as much force as I'd ever heard, "Jim, that's not mathematics". The storm passed, and we went out and had a cup of coffee afterward.

I had no idea why the idea upset him so; we never discussed it and remained friends until he passed away a couple of years ago. I even dedicated one of my books to him, as he was an inspirational teacher and, as I have said, perhaps the only mentor who could have successfully guided me on the path to a doctorate. He helped me obtain my teaching position, and even though we worked on many of the same problems, or ones that were closely related, during the 15 or so years after I received my doctorate, we never co-authored a paper. I would have enjoyed that immensely, and I think he would have, too.

But I knew he was wrong; the idea about "sharing the job" was mathematics, and it launched the second half of my mathematical career. It brought me into contact with a number of different collaborators in different countries – by now, email and the internet had revolutionized the way mathematics research was done by enabling mathematicians to communicate much more quickly, rather than having to confine such communication to snail mail, conferences, and visits.

The idea of "sharing the job" has not caught fire in the mathematical world, although it has prompted some research. But the idea is so incredibly universal, applicable to practically every one of the numerous branches of mathematics, that I'm sure it will eventually prompt some useful work. And even if it doesn't, it enabled me to do what I love to do – think about mathematics, and because it was an original idea, I was able to do productive work without having to play catch-up in a research area which mathematicians had already explored in some depth.

Finishing the Job

Over the next few years I obtained partial results, but the ultimate prize, the Banach Contraction Principle, where the load is shared by a finite number of functions $T, T^2, T^3, \ldots, T^N$, remained elusive. But then came a semester in which I was fortunate to have a prodigy in one of my math classes.

One of my colleagues was Kent Merryfield, an extremely talented mathematician who was more interested in learning mathematics and passing it on to aspiring young mathematicians than in doing research. One of the aspiring young mathematicians was his son, James. By the time James was 16, he had completed not only the standard high school math courses but all of lower-division college mathematics and some upper-division math courses as well. He enrolled in the second semester of my class in real analysis, and I could tell how talented he was from his first exam, although, of course, I knew it already from conversations with his father.

Most research in mathematics depends on knowing a lot of stuff. In order for me to do the research that ended up in my dissertation, I not only had to have finished several years of graduate school, but I had to read a number of papers that had only been published very recently. But such was not the case for the version of the Banach Contraction Principle on which I was working; my papers were the only ones in the area, and what you really needed to work on the problem was the ability to bring new ideas to it. And James had new ideas. It took about a semester, but between us we finished the problem and obtained the result for T, T^2, T^3, ..., T^N. When I say "between us", it's only fair to say that the vast majority of the new ideas that were needed came from James. James was – and I'm certain he still is – extraordinarily brilliant, but 16-year-olds are not the world's greatest communicators, and much of what I did was to clarify James' ideas. Fittingly, between us we "shared the load" of producing the desired result.

California State University, Long Beach, where I teach, is not a doctoral-granting institution, so I've never had the pleasure of having a student do research under my guidance, as I did research guided by Bill Bade. But I did have the inexpressible pleasure of working with an extremely talented young mathematician – and seeing him present our joint effort at a seminar at Cal Tech. As far as I know, not even Sheldon Cooper of *The Big Bang Theory* did that.

Ending the Quest

I retired from Cal State Long Beach in 2013, having written my last serious research paper in 2011. A few years after I retired, I received a request from a journal to referee an article. It was on fixed points, a subject very familiar to me, and several of my papers were cited in the bibliography. It should have taken me a couple of weeks to referee it. As I read the paper, I recognized arguments that I had used previously, but it was becoming more and more difficult to concentrate and to actually follow the arguments that were made. I think I did a decent job of refereeing, but it took me a couple of months, and when I finished the job, I realized that I was no longer capable of proceeding any further on this particular quest. It had provided me with two decades worth of satisfaction, but the time had come to pass the torch.

Most mathematicians – and this group definitely includes myself – don't make noteworthy contributions. The great results are the work of a few extremely talented mathematicians. But mathematics is perhaps the only edifice the human race has constructed that goes back more than two and a half millennia, and it is still ongoing. If we've placed a few bricks in the edifice, or even if we've simply enabled others to place those bricks, we at least have had the satisfaction of contributing to it.

My first quest – to solve the uniqueness of norm problem – had been unsuccessful, but my second quest with the Banach Contraction Principle had brought me everything that a good quest should. But, as I retired from Cal State Long Beach, I felt that my questing days were over. I didn't have any ideas in mind, and I wasn't capable of working on the problems that had interested me in the past.

But you never know what life has in store for you.

Chapter 11

The Multi-Armed Bandit Problem

Early in this book, I somewhat derisively remarked that the search for the best pizza in one's neighborhood does not qualify as a quest. I like pizza just as much as the next guy, and like the next guy, I tried a few of the neighborhood pizza parlors before settling on one. However, it wasn't a quest for two reasons: The search wasn't especially arduous and, hey, it's pizza.

A couple of years after I retired from California State University, Long Beach, my friend, Len Wapner, showed me an intriguing mathematical problem that he has named Blackwell's Bet. I'll describe it a little later; it was like nothing in mathematics I had ever seen. And it started me on a mathematical quest that has occupied me ever since, and it does relate to the search for the best pizza.

You probably haven't given a whole lot of thought to the search for the best pizza in your neighborhood. Maybe you have a friend who recommended a place and you liked it, so you continued to go there, or maybe you looked at a couple of reviews online. But it's a safe bet that you didn't regard this as a sufficiently absorbing problem to spend much time on it.

But if you consider the framework in which the search for the best pizza takes place, it becomes an absorbing mathematical problem. It goes by the name multi-armed bandit problem, and literally thousands of papers have been written about this. You can get some idea of the extent to which this problem has been investigated by taking a look at the following paper: https://arXiv.org/abs/1904.07272.

If you're not familiar with arXiv.org, just click on the link, and then on the right, click "View pdf". This paper has been revised a number of times since it was first written; it's now 188 pages long, and the last 20 pages contain the bibliography.

So, a brief description of the multi-armed bandit problem is in order. The name comes from the description of slot machines as one-armed bandits. Old-style mechanical slot machines featured a single mechanical "arm" that one pulled down after inserting a coin. The following link provides an example of what the display looked like if you want to try it. A word of caution: Play only the "play free" version because I'm guessing that the "play with real money" version is rigged against you (as were the original one-armed bandits): https://www.penny-slot-machines.com/free-slots.html.

There really isn't anything to interest a mathematician about this, but what if there are two or more arms, each having a different payout scale? The basic multi-armed bandit problem assumes such a framework and asks the question: Given a limited bankroll, what is the best way to play the multi-armed bandit to either determine the arm with the greatest payoff or to emerge with the greatest amount of money?

The problem of emerging with the greatest amount of money is more complex than the problem of simply discovering the best arm. That's because one must spend some of one's bankroll in attempting to discover the best arm (the exploration phase) and the remainder of one's bankroll in making money by playing the presumed best arm (the exploitation phase).

So, let's return to the pizza search. We needn't worry about the exploitation phase because, once we discover the best place for pizza, we're probably going back to it again and again until the quality of the pizza declines substantially – or until someone says that they love the pizza at a different establishment and we really should give it a try.

Merely defining the different types of multi-armed bandit problems is a serious exercise in mathematics, so we're going to confine ourselves to a few simple types. Because we started this discussion with pizza, we'll define those types in terms of pizza – at least in terms of quality.

One thing you could do is a comparison taste test: Order your basic mushroom-pepperoni pizza from two different places and blind-taste test each side by side. But who does that, unless you're either a food critic or a pizza fanatic? Most of us simply go into one, order a pizza, and eat it there or take it home.

So here you are, just having brought home a mushroom-pepperoni pizza and are about to bite into it. How are you going to evaluate it? Will you simply decide whether it's satisfactory, or will you assign it a number on a scale?

If you decide to assign it a number on a scale, you might be surprised to learn – as I was – that there's a way to have a better-than-fifty-percent chance of deciding which pizza is better, this one or the untasted one from the other restaurant, without tasting the other pizza! When I first learned about this, it blew me away – and so now it's time to discuss Blackwell's Bet.

I'll put the problem in the Blackwell's Bet format and then show you the parallel to the pizza problem, although there's a pretty good chance you'll see the parallel on your own. Let's say you have two envelopes, each with an undetermined amount of money in them. You are allowed to open one envelope, count the amount of money in it, and then make the following decision: either keep the money in the opened envelope or take the money in the unopened envelope.

We want to remove subjectivity from the decision, so let's assume that the amount of money isn't so small – say eight cents – that you'd unhesitatingly switch or so large – say ten thousand dollars – that you'd be tempted to keep it. So, let's call the amount of money in the opened envelope x.

Here's where a strategy suggested by the brilliant mathematician David Blackwell comes into play. Select a random number from somewhere – maybe the amount of your last utility bill. Anything will work. Call this random number r. If r is less than x, keep the amount of money in the opened envelope; otherwise, take the amount of money in the unopened envelope. Amazingly – at least to me the first time I heard it – this technique will give you greater than a fifty-percent chance of making the correct decision.

Here's why – and it's really mathematically pretty simple. Let p be the probability that the random number r is less than the smaller of the two amounts in the envelopes and q be the probability that r is greater than the larger amount.

The probability that you will have opened the envelope with the smaller amount is 1/2. In that case, the correct decision is to switch, and the random number r will tell you to do so with a probability of $1 - p$. The combined probability of having opened the envelope with the smaller amount and making the correct decision (to switch) is 1/2 $(1 - p)$.

The probability that you will have opened the envelope with the larger amount is 1/2. In that case, the correct decision is to keep the money, and the random number r will tell you to do so with a probability of $1 - q$. The combined probability of having opened the envelope with the smaller amount and making the correct decision (to keep the money) is $1/2(1 - q)$. Therefore, the probability of making the correct decision is 1/2 $(1 - p)$ + 1/2 $(1 - q) = 1/2 + 1/2\ (1 - (p + q))$.

$1 - (p+q)$ is the probability that the random number will be somewhere between the smaller and larger amounts of money, and that's always greater than or equal to zero. In fact, it's greater than zero except under very weird conditions (such as there was a penny in one of the envelopes and two cents in the other, and the random number you chose was a monetary amount). So, this strategy gives you better than a fifty–fifty chance of selecting the envelope with the larger amount.

I went through my entire mathematical career – more than 50 years – without ever hearing about this, and when I did, I was absolutely fascinated by it. Fascination is a big component of a quest, and ever since I retired from teaching, I've been looking at some of the problems for which this simple but surprising strategy can be used.

This is the third mathematical quest I have undertaken, and it differs substantially from the other two in that I'm embarking upon it while being seriously underequipped. I know almost no probability theory beyond what is taught in high school – well, maybe a little, but not much. So, I'm not expecting to accomplish anything memorable – at least, not something that will become part of the literature. When I was younger, I would never have considered embarking on such a quest. I found mathematical research rewarding not only because it put me in contact with truth (always) and beauty (surprisingly often) but also because there's a special thrill when you come up with a new result. For one moment, you are (probably) the only entity in the Universe in possession of this knowledge. That raises the stakes – the rewards are higher, but the prize is generally much more difficult to obtain. At any rate, you need a lot more in the way of mathematical tools to participate in mathematical research today, just as you need a lot more in the way of tools to participate in any form of research in which there are a large number of highly trained people.

David Blackwell once said, "I'm not interested in research, I'm interested in knowing the truth". I'm a big David Blackwell fan, as was almost everyone who knew him, but I think this is a little disingenuous. I've never met a mathematician or scientist who isn't interested in finding out something new that no one else knows, and that's certainly a large part of research.

I doubt that I'm going to find out anything new, as ill-equipped as I am for the task. But at this stage of my life, it doesn't matter; it will be new and fascinating to me. I could have included other chapters than this one in this book, but I especially wanted to include this one because the results are fairly easy to understand and some of them are surprising. But it's a lot easier to communicate the joy of a quest in which you have taken part, and my hope is that this chapter will motivate someone to learn what is needed to embark upon a quest of their own. Finding truth and beauty is a worthwhile goal for anyone, no matter in what area that truth and beauty interests you, even if you find that truth and beauty in mathematics.

Acceptable and Unacceptable Pizza

Let's say you've just moved into a neighborhood and are trying to find a good place for pizza. You find one nearby, order a pizza, and consume it, at which point you might ask yourself, "Would I order another pizza from this establishment?" You probably would if you found the pizza acceptable, and you probably wouldn't if the pizza was unacceptable.

You've just conducted what mathematicians call a "Bernoulli trial". A Bernoulli trial has two outcomes, which mathematicians label "success" and "failure". In this case, an acceptable pizza would qualify as a success and an unacceptable one as a failure. Moreover, subsequent trials are unaffected by the results of previous trials. If you order a second pizza from the same establishment, your taste buds won't care whether or not the first pizza that you ordered was acceptable.

There is only one parameter associated with a Bernoulli trial: its success probability. But there is an analog of Blackwell's Bet for Bernoulli trials. Suppose you are trying to decide which of two pizza establishments is more likely to produce an acceptable pizza, and you are only allowed to consume a single pizza from one of the two establishments. Let's call them Alberto's and Bernadino's. You can use the idea behind Blackwell's Bet to have a better than fifty–fifty chance of choosing the establishment that is more likely to produce an acceptable pizza.

Here's how you do it. Flip a fair coin to decide whether to order from Alberto's or Bernardino's. Let's say the result was that you order a pizza from Alberto's. If it is acceptable, guess that Alberto's has the higher probability of producing an acceptable pizza. If it is unacceptable, go with Bernardino's pizzas. The same mathematics as in Blackwell's Bet shows that you have a better than a fifty–fifty chance of choosing the establishment with better pizzas (at least, from your point of view). In fact, if p and q are the two probabilities of producing acceptable pizzas from the two different restaurants and $p > q$, this strategy gives you a probability of $1/2 + 1/2\,(p - q)$ of guessing the establishment producing more acceptable pizzas. And this is the best you can do if you're only allowed to order a single pizza.

But what if you could order two pizzas before reaching your decision? What's the best strategy for doing so – or is there even a best strategy?

Here's a straightforward idea. Order one pizza each from Alberto's and Bernardino's. If one is acceptable and one is unacceptable, choose the establishment that produced the acceptable pizza. If either both are acceptable or both are unacceptable, flip a coin.

This is obviously a very reasonable approach, so let's look at how it does. Let p and q be the probabilities that each establishment will supply an acceptable pizza, and assume that $p > q$. The mathematics here is pretty simple. We will guess correctly (by choosing the restaurant with the greater probability of producing an acceptable pizza) if we sample an acceptable pizza from the better restaurant (the one associated with p) and an unacceptable one from the poorer restaurant. This parlay occurs with a probability of $p(1 - q)$. If we receive two acceptable pizzas (occurring with a probability of pq) or two unacceptable pizzas (occurring with a probability of $(1 - p)(1 - q)$), we'll be forced to flip a coin.

So, the probability of making the correct guess is

$$p(1 - q) + 1/2(pq + (1 - p)(1 - q))$$
$$= p - pq + 1/2(2pq + 1 - p - q) = 1/2 + 1/2(p - q).$$

Well, that comes as a surprise – at least to me. The completely logical approach to selecting the better restaurant by ordering one from each yields exactly the same probability of success as if we'd just ordered a single pizza from one of the restaurants and used the Blackwell approach.

Maybe we'd do better if we flipped a coin to decide from which restaurant to order two pizzas. If both were acceptable, we'd obviously have greater assurance that we'd ordered pizzas from the better restaurant. And conversely, if both were unacceptable, we'd have good reason to go with the other restaurant. And if we received one acceptable pizza and one unacceptable pizza, we'd have three different options. The first is to flip a coin to choose between the two restaurants. The second would be to figure that one acceptable pizza is better than none and go with the restaurant from which we'd ordered the pizzas. The third is to figure that one unacceptable pizza is – well – unacceptable and go with the other restaurant.

Let's do the computations and see what happens.

Option 1: Flip a coin if we get one acceptable and one unacceptable pizza. With a probability of 1/2, we will order two pizzas from the restaurant with the better pizzas, and we will guess correctly if we get two acceptable pizzas. This has a combined probability of $1/2 \, p^2$. We will also guess correctly half the time if we get one acceptable and one unacceptable pizza, which has a combined probability of $1/2 \times 1/2 \times 2p(1 - p) = 1/2p(1 - p)$. So, the combined probability of both selecting the restaurant with the better pizzas and guessing correctly is $1/2p^2 + 1/2p(1 - p) = 1/2p$.

What if we order our pizzas from the restaurant with the inferior pizzas, which will occur with a probability of 1/2? We will guess correctly if we get two unacceptable pizzas, and we will guess correctly half the time if we get one acceptable and one unacceptable pizza. So, the probability of guessing correctly when we order from this restaurant is $1/2(1-q)^2 + 1/2q(1-q) = 1/2(1-q)$.

Therefore, the probability of guessing correctly if we order two pizzas from the same restaurant is $1/2p + 1/2(1-q) = 1/2(1+p-q)$. This is the third time we've seen this expression! So far, we have yet to encounter a strategy which gives us a different result, but we still have a lot of ground to cover.

Option 2: If we get one acceptable and one unacceptable pizza, guess that this restaurant is the better of the two (because one acceptable pizza is evidence that they can produce acceptable pizzas). With a probability of $1/2$, we will order two pizzas from the restaurant with the better pizzas, and we will guess correctly if we get at least one acceptable pizza. This has a combined probability of $1/2 \ p^2 + 1/2(2p(1-p)) = p - 1/2p^2$. With a probability of $1/2$, we will order two pizzas from the restaurant with the inferior pizzas, and we will only guess correctly if we receive two unacceptable pizzas. This occurs with a combined probability of $1/2(1-q)^2$.

So, the combined probability of guessing correctly using this guessing strategy is $p - 1/2p^2 + 1/2(1-q)^2 = 1/2 + p - q + 1/2q^2 - 1/2p^2$. This certainly looks different from the expression we've seen three times before, so let's do a little algebra. By rearranging the terms and factoring, we get the expression $1/2 + (p-q)(1 - 1/2(p+q))$, which is definitely a different expression from the one with which we're so familiar.

Let's do a little algebra to compare them. Suppose we wanted to know in which instances it was better to order one pizza from each establishment as a basis for making our decision than to order two pizzas from a single restaurant. We'd want to know when $1/2 + 1/2(p-q) > 1/2 + (p-q)(1 - 1/2(p+q))$. We can subtract $1/2$ from both sides and reduce it to the question of when $1/2(p-q) > (p-q)(1 - 1/2(p+q))$. Since $p > q$, we can divide both sides by the expression $p - q$ since it's always positive. This yields the inequality $1/2 > 1 - 1/2(p+q)$, or $p + q > 1$.

Now, what does this mean? In order to answer that, we have to examine what is meant by $p + q > 1$. It means that if we were to look at the two restaurants together, they'd be more likely to produce acceptable pizzas than unacceptable ones. You'd certainly feel that way if you were in the Little Italy section of town; the residents in that area undoubtedly know good pizza, and a poor pizza establishment would probably go out of business quickly. So, if you're in the Little Italy section of town, you're more likely to make the correct guess by ordering one pizza from each of the two restaurants than by ordering two pizzas from the same place.

Option 3: If we get one acceptable and one unacceptable pizza, guess that this restaurant is the poorer of the two (because one unacceptable pizza makes it more likely that you will be disappointed when you order one). Before proceeding with the analysis, here's an opportunity to test your intuition. In view of what you've already seen, what do you think we are going to be able to conclude from this assumption? I'm not talking about the actual formula that will result, but the interpretation that will come from looking at the formula.

With a probability of 1/2, we will order two pizzas from the restaurant with the better pizzas, and we will only guess correctly if we get two acceptable pizzas. This has a combined probability of $1/2 \, p^2$. With a probability of 1/2, we will order two pizzas from the restaurant with the inferior pizzas, and we will guess correctly if we receive at least one unacceptable pizza because our decision rule in this instance will force us to make the correct guess. The combined probability of choosing the inferior restaurant and getting at least one unacceptable pizza is $1/2((1 - q)^2 + 2q(1 - q)) = 1/2(1 - q^2)$. We could also have arrived at this expression by realizing that $1 - q^2$ is the probability of not getting two acceptable pizzas.

So, the probability of making a correct guess is $1/2 p^2 + 1/2(1 - q^2) = 1/2 + 1/2(p + q)(p - q)$. Again, we ask ourselves: If we use this rule, when is it better to order one pizza from each of two restaurants than to order two pizzas from the same place? That happens if $1/2 + 1/2(p - q) > 1/2 + 1/2(p+q)(p-q)$, and it's easy to see that this happens when $p+q < 1$. What this means is that it's better to choose two pizzas from one place if you're in a section of the city where the typical pizza you find is unacceptable by your standards. Maybe that's one in which there are very few pizza places, or the ones that you find are simply part of a large chain of pizza franchises.

You might think that our analysis is complete, but there's an entire class of strategies we have yet to consider. The ones we've looked at were determined in advance, that is, once we decided which rule to use, we went ahead and either ordered one pizza from each of the two restaurants or flipped a coin to choose the restaurant from which we would order two pizzas. What about the possibility of ordering one pizza from one of the restaurants and then making a decision on whether to order the second pizza from the same restaurant or switch and order the second pizza from the other restaurant?

Before doing the analysis, I felt that there might be an advantage to this approach. The strategies already discussed do not modify the strategy based on information received. In general, the best strategies do take such information into account - someone who doesn't take information into account is often referred to as pig-headed. So, it's certainly worth taking a look at doing so in this case.

The obvious strategy of this type would be to continue ordering from the same restaurant if the first pizza is acceptable, but switch to the other restaurant if the first pizza is unacceptable. That's what we do, but how should we guess depending on the results of our taste test?

Obviously, if we get two acceptable pizzas, we'll go with that restaurant – that's a no-brainer. If we get two unacceptable pizzas, we'll have to flip a coin. If the first pizza was unacceptable and the second pizza acceptable, we'd go with the second restaurant. But what if the first pizza was acceptable and the second wasn't? We have the same three options as we did in the case where we ordered two pizzas from the same restaurant and got one acceptable pizza and one unacceptable one

Let's do the math.

Case A: If the first pizza is acceptable and the second isn't, we flip a coin. With a probability of $1/2$, we'll order the first pizza from the better restaurant. In this case, we'll guess correctly if both pizzas are acceptable (probability: p^2) and half the time when the first pizza is acceptable but the second isn't (probability: $1/2 \times p(1-p)$) or both pizzas are unacceptable (probability: $1/2(1-p)(1-q)$. This yields a combined probability of a successful guess of $1/2(p^2 + 1/2p(1-p) + 1/2(1-p)(1-q)) = 1/2(1/2p^2 + 1/2 - 1/2q + 1/2pq) = 1/4(1 + p^2 - q + pq)$.

With a probability of 1/2, we'll order the first pizza from the worse restaurant. In this case, we'll guess correctly if the first pizza is unacceptable and the second one is (probability: $(1-q)p$) and half the time when the first pizza is acceptable but the second isn't (probability: $1/2 \times q(1-q)$) or both pizzas are unacceptable (probability: $1/2(1 - p)(1 - q)$. This yields a combined probability of a successful guess of $1/2((1 - q)p + 1/2q(1 - q) + 1/2(1 - p)(1 - q)) = 1/2(1/2 + 1/2p - 1/2q^2 - 1/2pq) = 1/4(1 + p - q^2 - pq) = 1/2 + 1/4(p - q)(1 + p + q)$.

We haven't seen this before, but it's clear when this works better than ordering one pizza from each of the restaurants – when $1/4(1+p+q) > 1/2$, or $p + q > 1$. *That* we've seen before – it's the Little Italy condition.

Case B: If the first pizza is acceptable and the second one is unacceptable, guess that this restaurant is the better of the two (because one acceptable pizza is evidence that they can produce acceptable pizzas). Hold on a moment! This looks like it may give a different condition because as soon as we receive an acceptable pizza, there's no need to order a second one since we know we're going to choose that restaurant.

With a probability of 1/2, we'll start by ordering a pizza from the better restaurant. If it's acceptable (probability: p), we'll choose that restaurant. If it isn't acceptable, we'll order a pizza from the other restaurant and will make the correct guess half the time if the pizza from the other restaurant isn't acceptable either (probability: $1/2(1 - p)(1 - q)$). So, we'll guess correctly in this case with a probability of $1/2(p + 1/2(1 - p)(1 - q)) = 1/2(1/2(1 + p - q + pq)) = 1/4(1 + p - q + pq)$.

With a probability of 1/2, we'll start by ordering a pizza from the inferior restaurant. We'll make the correct choice if it's unacceptable and the other pizza is acceptable (probability: $(1 - q)p$) or half the time if both pizzas are unacceptable (probability: $1/2(1 - p)(1 - q)$). So, we'll guess correctly in this case with a probability of $1/2((1 - q)p + 1/2(1 - p)(1 - q)) = 1/2(1/2p + 1/2 - 1/2q - 1/2pq) = 1/4(1 + p - q - pq)$.

Adding these two, we get the probability of making a correct guess as $1/2(1 + p - q)$. No different from ordering one from each restaurant – except with the probability of $1/2(p + q)$, we'll only have to buy one pizza to make our decision.

Case C: If the first pizza is acceptable and the second isn't, choose the other restaurant. At first glance, you might think this is the same as Option 3, but it isn't because Option 3 uses the criterion "one of two is acceptable" while this case is "first one is acceptable, the second isn't".

With a probability of 1/2, we'll start by ordering from the better restaurant. We guess correctly only if they produce two acceptable pizzas (probability: p^2). The probability of a correct guess is therefore $1/2p^2$.

With a probability of 1/2, we'll start by ordering a pizza from the inferior restaurant. We'll make the correct choice if either the first or second pizza is unacceptable, as, in either situation, we'd guess that the other restaurant is the better one. So, we'll guess correctly in this case with a probability of $1/2(1 - q^2)$.

Our probability of making a correct guess is therefore $1/2(p^2 + 1 - q^2) = 1/2(1 + (p + q)(p - q))$. This is superior to buying one pizza from each restaurant if $p + q > 1$ and inferior if $p + q < 1$.

Now, let's summarize and look beyond the mathematics because there are some very interesting conclusions. There are three different classes of strategies when there are two restaurants and we can buy two pizzas. The first class yields a probability which is identical to the one where we made our choice based on buying only one pizza. The second class yields a probability which is superior to the "single pizza purchase strategy" when $p + q > 1$ but inferior when $p + q < 1$. Finally, the third class is the opposite of the second class, yielding a superior probability when $p + q < 1$ but an inferior one when $p + q > 1$.

The first interesting conclusion is that if you don't know anything about the pizza restaurants in your neighborhood, you can save time and money by making your decision based on only one pizza. OK, since you'll probably be ordering pizza with some frequency, you probably won't save money, but if you're making a choice other than pizzas where each choice can be expensive, you *can* save money, time, or both by simply using the strategy associated with a single choice.

But this goes beyond pizzas – way beyond. Any time you have a choice between two alternatives, you may not need to spend the money and time investigating both and then agonizing over which of the two alternatives to pursue. Simply flip a coin to decide which to investigate, do that, and decide whether it is satisfactory or unsatisfactory. If it is satisfactory, take it; if not, take the other alternative.

Here's a simple example. Suppose you have decided to purchase one of two types of cars. Flip a coin to decide which of the two cars to investigate, then go to the dealer, check the price, and take a test drive. If you like it, buy it; if not, buy the other one.

In fact, any time you have a choice between two unknown options, this strategy is worth consideration. You need a good dentist, and you have two good friends who go to different dentists. Admittedly, since they're going to these dentists, there's a high probability that each is satisfied, but this strategy will help you choose the better of two options if one is satisfactory 90% of the time and the other 95% of the time. Flip a coin, call up one of your friends, and see how they feel about their dentist. If they're satisfied, go to that one. If they're considering switching, go to the other one.

Here's another common situation. Your daughter has narrowed the choice of colleges down to two. There's only time – or money – to visit one. Flip a coin to choose one of them to visit. If she likes it, she should go there. If not, go to the other one. Visiting both won't improve her chances of making the right choice, but you may have a hard time convincing her of that, unless she's planning to major in a subject requiring math..

You do need a certain amount of caution; I wouldn't recommend relying on this for life-altering decisions (see the third interesting conclusion for something else). Deciding which home to buy or which job to take is a little more significant, as well as nuanced, than buying a pizza – or even a car. But this policy can save you time, money, and aggravation on the less important decisions. You'd be right in the long run half the time if you simply flipped a coin, but you'll do better than a fifty–fifty chance if you follow the Blackwell strategy. And you don't need that second observation.

The second interesting conclusion is that if you do know something about the pizza restaurants in your neighborhood, you can obtain value from the second pizza by using a strategy which is superior either when $p + q > 1$ (a good neighborhood for pizza) or $p + q < 1$ (a bad neighborhood for pizza). And this, too, has important economic consequences. Suppose you want to use the most reliable of two components for your product, and in general those components are pretty reliable. You'd go with one of the strategies which is superior in the case of $p + q > 1$. And conversely, if you need to choose between two subpar choices, use one of the strategies which is superior in the case of $p + q < 1$.

The third interesting conclusion is that there may be a tendency for someone to say, "Doesn't this analysis show that I don't really need to get a second opinion on whether I should have surgery?" That's an incorrect application of these results. With the exception of the case where you decided in advance to order one pizza from each restaurant (and that's like getting two different opinions from two different doctors), every analysis began with a fifty–fifty choice of which restaurant from which you either ordered two pizzas or from which you ordered the first pizza. That fifty–fifty choice is necessary for the mathematical analysis to work. However, if you go to your internist or specialist first, you aren't making a fifty–fifty choice because you're always seeing *your* doctor first.

Although I've done a lot of additional work on this problem, I probably haven't made any notable breakthroughs. It's a well-studied problem by people vastly more knowledgeable in this particular area than I am. But one of the features of a good quest is that it obsesses you; it gives you a reason to get up in the morning, hoping you'll discover something interesting. As David Blackwell said, he wasn't interested in doing research, he was interested in finding out the way things were. I spend some of my time reading about recent developments in science because it's been an area of interest to me all my life. There are many enthralling quests that are currently taking place in science, and I enjoy reading about them. But I can't think about them because I don't have the tools to do so. Though I only have a few tools – the elementary ones I've used in this chapter – I can continue to think about problems in this area. And having fascinating problems to think about has been one of my main sources of enjoyment for most of my life.

Chapter 12

CHaikus

When found together
The true and the beautiful
Enchant and delight.

I married for the first and only time when I was in my late fifties, and my wife and I decided not to have children, as we were the sole source of support for her mother and sister. But I've always enjoyed working with children and have had the good fortune to mentor some very bright ones. Two of the children I mentored were the twin sons of a couple who lived in Beverly Hills. At one point, their mother said something that made a lot of sense to me. Everyone is given one hundred points at birth; they're just distributed differently for each person.

As you might gather, the majority of points that I was given were in math and science, but surprisingly, I've always been able to write verse. I think of verse as poetry designed to amuse rather than provoke deep thought. I present as an example a verse I composed many years ago:

The White-Knuckled Flyer

I'm not an air-traffic controller,
Nor am I a part of the crew,
But equal to theirs in importance,
Is much of the work that I do.
My job, as a white-knuckled flyer,
Whenever there's wind or there's rain,
Is holding on tight to the armrests,
In order to hold up the plane.

> I never sit next to a window,
> I need to use both of my hands,
> I must be prepared in an instant,
> To hold up the plane 'til it lands.
> While flying through turbulent weather,
> My shoulders and forearms will ache,
> Quite often I've held on for hours,
> With nary a five-minute break.
>
> And when, if you're sitting beside me,
> The plane gives a shuddering jerk,
> Don't worry about my complexion,
> Though white-faced with fear I'm at work.
> So go have yourself a martini,
> And know that you haven't a care,
> Your seatmate's preventing disaster,
> By holding the plane in the air.

So, though I'm mostly a nerd, I do have other abilities. And we do have the TV show, *The Big Bang Theory*, to thank for an important contribution to the general understanding of nerds. Like me, nerds are interested in other things besides the subject (think math and science) that get them qualified as nerds. Sheldon Cooper, the central character, was interested in – among other things – trains, flags, and superheroes.

Although, now that I think of it, there are some nerds who are almost exclusively nerds. Paul Ehrlich was one of the great contributors to human welfare. After nearly six years of testing, he discovered a chemical compound that cured syphilis – and in so doing, created chemotherapy. So great was the respect in which Ehrlich was held that when he died during World War I, both sides used the occasion to pay homage to him.

However, Ehrlich was a nerd in the classic sense. When he went to a restaurant, he would often scribble chemical formulas on the linen tablecloths (this was in an era before paper tablecloths). Undoubtedly he embarrassed his dining companions.

I'm neither Sheldon Cooper nor Paul Ehrlich, but I do have outside interests. I've discussed some of them in this book, but I've always been fascinated by poetry. Many languages feature some form of poetry. I believe that in order to be classified as poetry, the work has to have some formal structure. I must admit I'm a big fan of both meter and rhyme; for me, the best poems feature both. And if you can weave in something else, such as deep thoughts or onomatopoeia, so much the better.

I first came across haikus in a James Bond novel, *You Only Live Twice*, some of which takes place in Japan. Bond is asked to compose a haiku – a three-line composition with five syllables in the first line, seven in the second, and five in the third – and he musters up:

> You only live twice,
> Once when you're born, once when you
> Look death in the face.

I'd give him a B for this effort. Interesting idea, but there's that incomplete (from a sentence standpoint) second line, which for me is a significant distraction.

Those familiar with developments in high-energy physics must be aware that new particles are often produced by slamming together existing particles at colossal speeds. I was surprised to learn that new poetic forms can be produced by slamming together apparently unrelated thoughts at any speed. It took me more than half a century to slam two apparently unrelated ideas together and, in so doing, produce the poetic form that I call a "CHaiku".

The first apparently unrelated idea came indirectly from Mike Lawrence, a world champion bridge expert whom I've known ever since my Berkeley days. One day, recently, I recalled a bridge hand that had been shown to me more than half a century ago by Mike Lawrence. He asked me what I would bid on the hand and, on hearing my response, told me that the purpose of the question was to determine in which portion of the country the respondent learned how to play bridge.

A bridge hand contains 13 cards, which are presented by suits – spades, hearts, diamonds, and clubs. Mike presented the hand in the following form (you don't need to know how to play bridge, so I've used Xs to denote cards):

$$
\begin{array}{cccc}
X & X & X & X \\
X & X & X & \\
X & X & X & \\
X & X & X &
\end{array}
$$

Someone who learned bridge on the West Coast sees the hand as presented horizontally – one suit of four cards and three suits of three cards each. Someone who learned bridge on the East Coast sees the hand as presented vertically – three suits of four cards each and one suit of one card. The different shapes elicit different bids, and by this bid, Mike was able to discern where the respondent learned how to play.

The second idea is indirectly the result of my wife's being Taiwanese. For a time, her mother lived with us, and she had some very traditional Chinese ideas with respect to home decorations. One was a scroll, which I frequently observe, as it is strategically located. I've shown a picture of it in Fig. 12.1.

Fig. 12.1.

I suspect that it has deep significance, but I think of it as "The Duck" because that's what the large central character looks like to me. As you probably know, a feature of the Chinese language is that, unlike English, it is read top to bottom and right to left.

And now to the CHaiku. A CHaiku is a haiku that fits into a three-by-three matrix (think rows 1–3 and columns A–C of an Excel spreadsheet) such that it not only reads as a traditional haiku (left to right and top to bottom) but also as a haiku if read as if written in Chinese – top to bottom and right to left.

It took me some time to construct my first CHaiku. It's not so great, but I present it here for its historical value – if any:

	Column 1	Column 2	Column 3
Row 1	poets	and	women
Row 2	who	truly idolize	love
Row 3	beguile	some	children

You're probably comfortable reading it left to right, top to bottom, but let me present the Chinese right-to-left version:

Women love children
And truly idolize some
Poets who beguile.

Well, the first version of a mathematical proof is sometimes awkward, but time and the efforts of other mathematicians often whittle it down to *a gem of purest ray serene*. Hopefully, the same will happen here.

Any new poetic form awaits the verdict of the literary public. Will CHaikus rise briefly to prominence, like the Tom Swifties of the 1960s, only to sink into a justly merited obscurity? Or will they take their place beside the classical haiku and iambic pentameter as a legitimate verse form? Only time will tell.

CHaikus can be represented by a three-by-three matrix with the number of syllables as entries in each location. The CHaiku presented above has the following matrix form:

$$2 \quad 1 \quad 2$$
$$1 \quad 5 \quad 1$$
$$2 \quad 1 \quad 2$$

The total of each of the first and third rows is 5, and the total of the second row is 7 – a classical haiku. The total of each of the first and third columns is 5, and the total of the second column is 7.

It is an interesting problem to list all the different matrix types of CHaikus (there aren't that many), and this leads to the following.

Table 12.1.

1	1	3
1	2	2
1	3	1
2	1	2
2	2	1
3	1	1

The Fundamental CHaiku Classification Theorem: There are 26 different matrix structures for CHaikus.

See! There's math everywhere – even in CHaikus!

If we consider a CHaiku in matrix form, it has three rows and three columns. Using the usual matrix notation of a_{ij} denoting the entry in the ith row and jth column, the structural requirements for a CHaiku are as follows:

(1) All the entries must be positive integers.
(2) The first and third row as well as the first and third column must sum to 5.
(3) The second row and the second column must sum to 7.

At any rate, this is simply a matter of counting – and relatively unsophisticated counting at that. We can make things easier for ourselves by realizing that the requirement that all entries be positive integers ensures that no entry in the first or third row or the first or third column can be more than 3 and that a_{22} cannot be more than 5.

If we list all the possible rows (and columns) that total 5 with all non-zero entries that sum to 5, we get Table 12.1.

If we list all the possible rows (and columns) that total 7 with all non-zero entries that sum to 7, with no number larger than 3 in the first or third position, we get Table 12.2.

So, it's simply a matter of selecting two choices from the sum-to-five possibilities and one from the sum-to-seven possibilities and then seeing whether it's a legitimate haiku form. This doesn't need to be a computer proof, as there are only $6 \times 6 \times 9 = 324$ different possibilities.

Table 12.2.

1	3	3
1	4	2
1	5	1
2	2	3
2	3	2
2	4	1
3	1	3
3	2	2
3	3	1

But we can reduce things fairly quickly. For instance, if we select 1 1 3 as the form for the top row, the third column must have the form 3 1 1. The matrix now looks like

$$\begin{matrix} 1 & 1 & 3 \\ a_{21} & a_{22} & 1 \\ a_{31} & a_{32} & 1 \end{matrix}$$

As soon as we specify a_{21}, all the other numbers are determined (or not) as follows. If we let $a_{21} = 1$, then $a_{22} = 5$ because the second row must sum to 7, $a_{31} = 3$ because the first column must sum to 5, and then $a_{32} = 1$ because the third row must sum to 5. Let's see if this assignment yields a legitimate CHaiku:

$$\begin{matrix} 1 & 1 & 3 \\ 1 & 5 & 1 \\ 3 & 1 & 1 \end{matrix}$$

Looks good! Row sums and column sums are what they should be.

How about if we let $a_{21} = 2$? Then, $a_{22} = 4$, $a_{31} = 2$, and $a_{32} = 2$. This gives

$$\begin{matrix} 1 & 1 & 3 \\ 2 & 4 & 1 \\ 2 & 2 & 1 \end{matrix}$$

Also OK.

What about if we let $a_{21} = 3$? Then, $a_{22} = 3$, $a_{31} = 1$, and $a_{32} = 1$. We get

$$
\begin{array}{ccc}
1 & 1 & 3 \\
3 & 3 & 1 \\
1 & 3 & 1
\end{array}
$$

This is clearly a good strategy to follow, but every so often it runs into an assignment which won't work. For instance, suppose that the top row is 1 2 2. If we let $a_{23} = 1$, this gives the outline

$$
\begin{array}{ccc}
1 & 2 & 2 \\
a_{21} & a_{22} & 1 \\
a_{31} & a_{32} & 2
\end{array}
$$

Our possible choices for a_{21} are 1, 2, and 3. If $a_{21} = 1$, then $a_{22} = 5$ in order that the second row sums to 7. But since $a_{12} + a_{22} = 7$, the second column must either have $a_{32} = 0$ (not allowed) or the second column totals more than 7 (also not allowed).

Sometimes all you can do is just work through the cases. I did so using a simple Excel program, in which I just listed the various possibilities for Rows 1 and 2, computed what Row 3 had to be from the various column requirements, and saw whether it satisfied the CHaiku form requirements. Interestingly, there are 26 possible structures – the same as the number of letters in the alphabet. Coincidence – or hidden forces at work? I suspect it's only coincidence, but how can you not love the fact that a poetic form having both Japanese and Chinese components has precisely the same number of possible forms as the number of letters in an alphabet that is the foundation of many Western languages?

At any rate, this enables a classification of the various acceptable matrix forms using the letters of the alphabet; I have done so in Table 12.3.

One of the projects on my bucket list is to compose a CHaiku for every one of the allowed forms. However, at this point, mathematical curiosity triumphed over literary aspirations, as I wanted to see what more would be possible in the way of extending haiku forms.

Table 12.3.

Type	Row 1			Row 2			Row 3		
A	1	1	3	1	5	1	3	1	1
B	1	1	3	2	4	1	2	2	1
C	1	1	3	3	3	1	1	3	1
D	1	2	2	1	4	2	3	1	1
E	1	2	2	2	3	2	2	2	1
F	1	2	2	2	4	1	2	1	2
G	1	2	2	3	2	2	1	3	1
H	1	2	2	3	3	1	1	2	2
I	1	3	1	1	3	3	3	1	1
J	1	3	1	2	2	3	2	2	1
K	1	3	1	2	3	2	2	1	2
L	1	3	1	3	1	3	1	3	1
M	1	3	1	3	2	2	1	2	2
N	1	3	1	3	3	1	1	1	3
O	2	1	2	1	4	2	2	2	1
P	2	1	2	1	5	1	2	1	2
Q	2	1	2	2	3	2	1	3	1
R	2	1	2	2	4	1	1	2	2
S	2	2	1	1	3	3	2	2	1
T	2	2	1	1	4	2	2	1	2
U	2	2	1	2	2	3	1	3	1
V	2	2	1	2	3	2	1	2	2
W	2	2	1	2	4	1	1	1	3
X	3	1	1	1	3	3	1	3	1
Y	3	1	1	1	4	2	1	2	2
Z	3	1	1	1	5	1	1	1	3

What About Haikubes?

It's like they said in *Star Wars*: Once you set foot down a dark path, it's hard to turn back. Well, maybe they didn't say that, but they probably said something like that. So, having come up with the Fundamental Classification Theorem for CHaikus, I wondered whether or not it would be possible to create other haiku variations using ideas from mathematics.

The first thing that occurred to me was to extend haikus into three dimensions – creating, as it were, a haikube. But what would such an object look like? Of course, one could take a cube and place a haiku on each one of the six faces: top and bottom, left and right, and front and back. But that lacks the feature of interconnectedness that characterizes the CHaiku.

So, perhaps cubes weren't the right object to use to create a three-dimensional structure that supports interconnected haikus. And what exactly could one do to interconnect haikus?

One possibility would be to continue the 5–7–5 pattern that characterizes the haiku by creating a cyclic haiku chain. A cyclic haiku chain of length 6 would consist of 6 lines. The odd-numbered lines would each contain 5 syllables, whereas the even numbered lines would each contain 7 syllables. Each line of 5 syllables would start a haiku, and to make the structure cyclic, the last line of the haiku that began with line 5 should be the first line of the haiku that begins with line 1. It's probably easier if I demonstrate with an example:

Line #	Line
1	When found together
2	The true and the beautiful
3	Enchant and delight
4	Quintessential endeavors
5	Of lasting value
6	Are romance and excitement

Besides the haiku that introduced the chapter, the other two are lines 3–5 and lines 5, 6, and 1.

We can create a three-dimensional structure by writing lines 1–6 in successive lines on a piece of paper and joining the bottom of the paper to the top, creating what I call a "haiku bracelet". Although this is a mathematical structure (not surprisingly, a cylinder), it isn't really all that fascinating.

But suppose we introduce the requirement that each haiku of the haiku bracelet fit into one of the 26 CHaiku matrix structures. We've loosened the requirement for it to be a CHaiku, so the vertical (Chinese) reading of the haiku doesn't need to make sense, but otherwise it retains the structural requirements. Mathematicians do this all the time – if they can't solve a problem with a given set of hypotheses, they see how much they could weaken the hypotheses in order to obtain a solution.

Let me give an example. Suppose that lines 1–3 of the haiku bracelet consisted of the following lines:

	Column 1	Column 2	Column 3
Row 1	Ending	this	volume
Row 2	A	very challenging	task
Row 3	But one	I	will do

This is a type-P (weak) CHaiku. The third line above must be the first line of another (weak) CHaiku, so the only possibilities are types O, P, Q, and R. But the third line of the second (weak) CHaiku must be the first line of the third (weak) CHaiku, and the third line of that (weak) CHaiku is the first line above, which has the 2–1–2 pattern.

A type-O haiku has the third line pattern 2–2–1, so we need to see if there is a CHaiku type whose first line has pattern 2–2–1 and whose last line has pattern 2–1–2. And indeed there is – type T. So, this leads us to believe there is a type-POT bracelet composed of weak CHaikus. Let's check:

Line #	Syllable Structure
1	2–1–2
2	1–5–1
3	2–1–2
4	1–4–2
5	2–2–1
6	1–4–2

The obvious first question to ask is – how many different CHaiku bracelets of length 6 are there? But that's just where a mathematician would start. Other questions that quickly arise from this one are whether there is a general formula that would give the number of different CHaiku bracelets of length $2n$, and what is the longest CHaiku bracelet that can be formed without repeating CHaiku types.

The Science of Patterns

One of the descriptions of mathematics is that it is the science of patterns. One never knows where patterns will arise; they can come from anywhere at all. I had gone my entire life without thinking about the possible patterns involved in creating haikus and extensions of haikus, but I'm willing to bet I'm not the first to wander down this path (although I think I'm the first to consider CHaikus). And sometimes these paths lead nowhere, but sometimes the investigation of patterns leads to something of substantial value.

I described earlier how, while at Cornell University, the great physicist Richard Feynman saw a plate with a pattern on it tossed in the air. As the plate spun, the pattern on it traced out a trajectory in time and space. Feynman decided to investigate that particular pattern – and doing this led him directly to the discovery of equations underlying quantum electrodynamics. Feynman obviously had no idea that this would happen; he just saw something he considered beautiful (the patterns on the spinning plate) and wanted to use the mathematical tools available to him to see where it would lead.

And that's the charm of a mathematical quest. Sometimes it's an end in itself, but sometimes it leads to something of enduring truth and beauty. You never know – but you'll never know unless you take the first steps down the road.

Index